The Triple Witching Hour

Essays, 2011-12

by David R. Roell

Astrology Classics
Bel Air

Celebrity photos courtesy Wikimedia Commons.

On the cover:
Le bas côté nord from St. Germain des Prés, Paris, 1981.
Photo by David R. Roell.
Format 6 x 6 cm, Ektachrome 64, Minolta Autocord, 10 second exposure.

ISBN: 978 1 933303 47 5

BLUE MOON EDITION
31 August 2012

Copyright © 2012 by William R. Roell.
All rights reserved.

Published by
Astrology Classics

The publication division of
The Astrology Center of America
207 Victory Lane, Bel Air MD 21014

On the net at www.AstroAmerica.com

🕓 Table of Contents

Introduction vii

Occupy Wall Street, *part 1* 1
Occupy Wall Street, *part 2:* Time and Time Again 10
The chart of the week: Howie Mandel 19
Losing Your Virginity 28
Charles Baudelaire: How to Read a Stellium 34
Daffodils and haircuts 40
William Shatner: Sun and Moon 41
Whole Sign Houses 47
Newt Gingrich Hates Mitt Romney 48
Hellenists, Medievalists 57
Whole Sign Houses 58
   Vettius Valens calculates the ascendant 58
The Florida Primary 67
Madonna 69
Lady Gaga 75
George McCormack's Weather Astrology 84
Whitney Houston 85
Michael Jackson 93
Daffodils 100
Rush Limbaugh 101
The Heat Wave 110
The Spring Equinox Chart 111

AstroMeteorology and Astro*Carto*Graphy 114
Ludwig 115
   Mozart's Birth and Death 123
   The Program to the Eroica 129
   The Program to the Ninth 131
Spring Planting 134
Know your leaders: Antonin Scalia 135
   Table of Alchochoden 141
Earthquakes 142
My Dinner with Andrea 142
Know your leaders: Joseph Biden 143
Excerpt from Long Range Weather Forecasting 149
Astrology Under Our Feet 150
People as Transits 161
Introducing Luke Broughton 165
Everybody's favorite host: Tom Bergeron 167
Venus, Money and the Eclipse 173
The New Medicine 174
The Development of Science 180
Dorothy and Her Magic Broomstick 184
Tropical and Sidereal: Clint Eastwood 189
Fix Your Heart with Astrology and Herbs 200

Bibliography: Books referenced in this book 202
About the Author 204

⏱ Table of Contents, *by subject*

I. *Celebrities and people:*
Baudelaire, Charles 34
Beethoven, Ludwig van 115
   Mozart's Birth and Death 123
   The Program to the Eroica 129
   The Program to the Ninth 131
Bergeron, Tom 167
Biden, Joesph 143
Eastwood, Clint 189
Houston, Whitney 85
Jackson, Michael 93
Limbaugh, Rush 101
Lady Gaga 75
Madonna 69
Mandel, Howie 19
Scalia, Antonin 135
Shatner, William 41

II. *Politics and the 2012 Election:*
Newt Gingrich Hates Mitt Romney 48
Florida Primary, The 67
Occupy Wall Street, *part 1* 1
Occupy Wall Street, *part 2:* Time and Time Again 10
Venus, Money and the Eclipse 173

continued:

Table of Contents by subject, *continued*

III. *AstroMeteorology:*
AstroMeteorology and Astro*Carto*Graphy 114
Daffodils 100
Daffodils and haircuts 40
Earthquakes 142
Excerpt from Long Range Weather Forecasting 149
George McCormack's Weather Astrology 84
Heat Wave, The 110
Spring Equinox Chart, The 111
Spring Planting 134

IV. *Astrology as an Earth Science,* also **Medicine**
Astrology Under Our Feet 150
Development of Science, The 180
Dorothy and Her Magic Broomstick 184
Fix Your Heart with Astrology and Herbs 200
New Medicine, The 174

V. *Astrology and Miscellany*
Hellenists, Medievalists 57
Introducing Luke Broughton 165
Losing Your Virginity 28
My Dinner with Andrea 142
People as Transits 161
Table of Alchochoden 141
Whole Sign Houses 47
Whole Sign Houses 58
    Vettius Valens calculates the ascendant 58

☉ Introduction to the third book of essays

As with the previous volumes, material in this book was taken from my weekly newsletters and covers a variety of topics. In addition to the usual celebrities, politicians and current events, there are two main themes:

First, that what we know as astrology are in fact energies generated by the Earth itself. As this is the inversion of what astrologers have believed for many thousands of years (*astrology is sky-based*), the concept came to me slowly, and with a great deal of struggle. In this book you will see some of the last bits of that struggle, as I have suppressed earlier guesswork.

But it was not enough to explain astrology. It was also necessary to understand how modern science could persist in denying what was overwhelmingly obvious. This was not as hard, as it was essentially tracing the results of the Twelfth Century Translators, first to Italy, then to Germany, and then to note the curious role of France. Who refused to concern itself with Renaissance science, but who were subsequently eager to impose a French dictat upon a defeated Germany after the 30 Years War, which we know today as The Enlightenment.

The second theme, more briefly stated, is my life-long quest to understand the relationship between Wolfgang Mozart and Ludwig van Beethoven. A proposed rectification of Beethoven's chart was an excuse to show part of my work in this area. As I caution at the start, the biography of Mozart that emerges is deeply radical.

As I find time I want to expand both of these topics into proper books. For the meanwhile, what is published here must do. *Enjoy!*

David R. Roell
August 27, 2012

🕐 Legend

Signs			Planets	
0°	♈	Aries	☉	Sun
30°	♉	Taurus	☽	Moon
60°	♊	Gemini	☿	Mercury
90°	♋	Cancer	♀	Venus
120°	♌	Leo	♂	Mars
150°	♍	Virgo	♃	Jupiter
180°	♎	Libra	♄	Saturn
210°	♏	Scorpio	♅	Uranus
240°	♐	Sagittarius	♆	Neptune
270°	♑	Capricorn	♇	Pluto
300°	♒	Aquarius	☊	Node (north)
330°	♓	Pisces	☋	Node (south)

☉ Occupy Wall Street

Occupy Wall Street was a short-lived American protest movement. It was suppressed. I include this essay here as it is an interesting example of mundane chart delineation. — Dave

I've had several requests for my thoughts on the Occupy Wall Street protests, which have been underway since September 17, 2011, at 12:01 pm in lower Manhattan. I put it off as I did not think the chart would read clearly. To my dismay, the chart is very clear.

Yes, there is a T-square, a tight one, in cardinal signs. If this is where you start in a chart, you will not have the proper context and so will, most likely, attribute the T-square to the protesters and then wonder why your analysis fizzles out.

The first task is to establish what kind of chart this is. For natal charts, we look at the Sun, the Moon, the Ascendant, and finally the Ruler of the Ascendant. We look at their signs, we look at their houses, we look at their rulers and we find the thread that connects them all to the person in front of us. Along the way we pick out relevant aspects, as well as aspect patterns, such as T-squares.

Occupy Wall Street is not a chart for an individual. It is the chart of a movement. This movement has a goal, or more precisely, a *target*, since it is defined as a protest. Which is adversarial, which makes it a form of warfare, or contest, or, in sporting matters, a game.

<u>For charts of conflict</u>:

The Ascendant and its ruler are given to the home team, in this case, the protesters.

The Descendant and its ruler are given to the challengers, in this case, Wall Street.

It is then a simple matter to judge which party is stronger. This is the traditional rule. You will find it in any good book on horary, such as those by Anthony Louis (my personal favorite), William Lilly, Derek

Appleby, etc. John Frawley has written an entire book on just this narrow subject, the brilliant *Sports Astrology*. I tell you this now so that you will know where I am going. My methods will not quite be horary, but rather, will be similar to those of Jean Baptiste Morin. Why do I not cite Morin himself? Because, to my wonder, Morin, a good French Catholic, saw horary as an infidel Islamic plot (Book 26) and so would have nothing to do with it.

Having established the *nature* of the chart, the next job is to establish its *validity*.

This can be done by various means, but the most simple is the best.

For Occupy Wall Street, we find Scorpio rising. The ruler of Scorpio is Mars. We find it in Cancer, late in the 8th house.

The 8th is, among other things, the House of Other People's Money. Other people's money is what Wall Street trades in, it is the very essence of their existence. The chart is therefore valid and can be judged. Preliminaries having been disposed of, we now have combat.

In horary, we read individual degrees. We start with the ascendant, where we find 27 degrees of Scorpio, a late, mature degree. We judge that while the movement will have time-tested experience, it will also be late, weak, feeble, elderly and worn out. That Scorpio is fixed as well as the most intense sign, we consider that age may have made it bitter and brittle, and that it may think that "everybody knows" what the problem is, so that they do not need to actually state it. In other words, our analysis has already explained why there is no list of demands.

The ruler of Scorpio is Mars. We find it at the very last degree of Cancer. At the very last degree, it has literally "run out of room" — and time, it is "painted into a corner," it has "no where to go," it is desperate, at the end of its rope, it has nothing to lose. That it is also at the very end of the 8th house shows that it has been here a long time, that it has had many adventures. With Cancer intercepted in the 8th, we realize the protesters were never in control (i.e., Cancer has no house cusp to claim as its own) and were played as suckers by Wall Street.

Considering this late 8th house Mars is the ruler of a protest chart, we may presume that it has, consequently, "gone down fighting" and as a result has already lost all its money. Thus the reason for the protest.

Where did the money go? To "follow the money," we need only follow the chain of dispositors.

Cancer is ruled by the Moon. We find the Moon in Taurus in the 6th house. As we find Taurus on the 7th house cusp as well it is easy to presume we have a 7th house Moon and give it to the other side, Wall Street. Which gets us where we want to go, but obscures useful details.

I have established there is no orb on the back side of an angle. Instead of placing the Moon in the 7th, we note her to be in Wall Street's 12th house, the house of secrets and plots. (The 6th house is the 12th from the 7th: Chart turning.) Wall Street keeps its true feelings carefully hidden. As the sign in question is Taurus, and as Taurus relates to money, we could say that Wall Street keeps its own proper money carefully hidden. Which we know to be true. When Hollywood has a big year, all of Los Angeles celebrates. With few exceptions, Wall Street always has a good year, but it never seems to make any difference to New York. Where do those lavish Xmas bonuses go? Note also that the Moon is exalted in Taurus. We will see more of this in a moment.

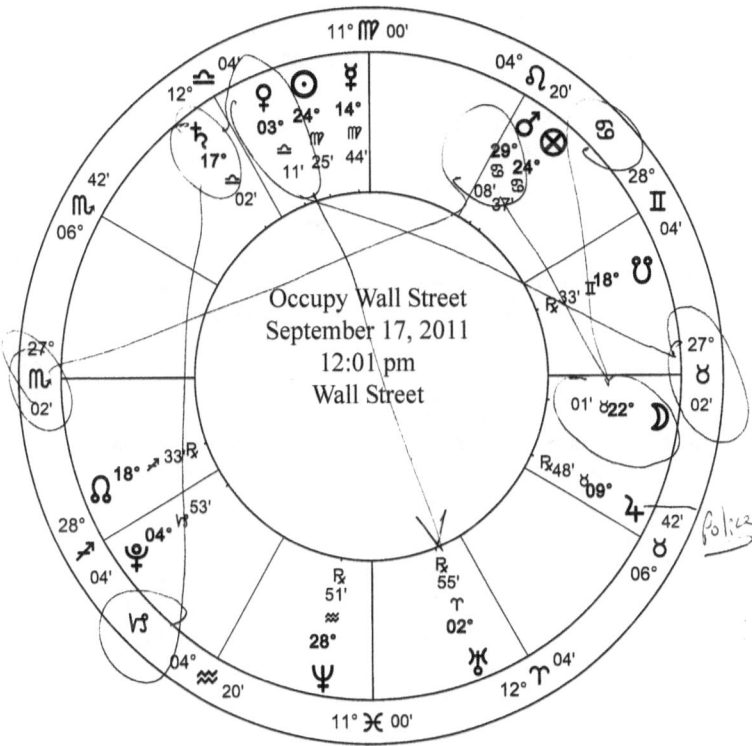

Both the Moon and the descendant are ruled by Venus. We find Venus in Libra, which it rules, and while it looks to be in the 10th, of publicity, it wants to be in the 11th, of friends, and whatever you think of a planet "wanting to be in that house," note that Venus herself rules the 11th house cusp. The friends we find there are hers and hers alone.

With Venus we have arrived at Wall Street itself. We find Venus at a mere 3 degrees, and while this is youthful and inexperienced, since Venus rules the sign, its instincts will generally be right and it will learn quickly. This is in sharp contrast to the Mars of the protesters, which we found to be old, worn out, and a mere tenant living in someone else's house (sign). Venus in its own sign is its own mistress, she, unlike Mars, not only answers to no one, but expects Mars, by way of her friend the Moon, to answer to her.

Do you begin to understand, already, what a hard job the protesters have set for themselves? I regret it gets worse.

In a mundane chart, which this also is, the 6th house is given to the army. In this case, the militarized police of New York, Los Angeles, Denver, Boston, Oakland, etc.

We find Taurus on the cusp of the 6th. Taurus is ruled by Venus, which means the police are under her command. Jupiter in the 6th in Taurus, Venus has a lot of police at her disposal, and they, money-lovers all, are well paid.

Jupiter rules Sagittarius. In Sagittarius we find the second house cusp. Which represents the money and resources of the protesters. We note the police (Taurus, 6th) have repeatedly destroyed protestor property (2nd, Sagittarius). In New York, and I believe elsewhere, there was a story that the property of the protesters had been "removed" and could be "claimed," but the inconjunction between the signs of Sagittarius and Taurus meant this story, even if well-meaning, was not true, and that the property had been taken to the dump. Beyond reach. Like a Sagittarian arrow shot far, far away.

Jupiter also rules the 4th, Pisces. The 4th represents land, specifically, the parks the protesters camp in. In a night chart Mars is said to be co-ruler of all three water signs, which, if this was a night chart, would give the protesters standing in the various parks. Instead, Jupiter — the police — have attacked and dispossessed the protesters in the dead of night.

Take a moment and get a breath. I started with the protesters, ruled by the ascendant, Scorpio, and its ruler, Mars. Which immediately led me to Wall Street, where I found the Moon, and then Venus, and now Jupiter, and I am, *alas!*, hardly done with Wall Street.

Drag yourself back to the 11th house. In it you will find Saturn in Libra. With exactly two exceptions, Venus hates Saturn. Well, what's to love about Saturn? He is sour, unfriendly, negative, nasty.

Except when Saturn is in Libra and Venus is with it in Libra, or when Saturn is in Libra and Venus is in Saturn's sign of Capricorn. In these two cases, the two are lovers, joined at the hip.

With Saturn in the 11th, ruled by Venus, Venus has few friends. But, Saturn representing the government, and Saturn in Libra specifically representing the courts and law, the friends Venus has are the ones that count. Important friends. Powerful friends. Note these should be the friends of Mars, not Venus, since Mars is the ruler of the chart as a whole, and the 11th house is the 11th house *from the ascendant.* Venus — Wall Street — has powerfully usurped Mars.

Venus's proper friends would be in the 11th house from the 7th, which is the 5th. Where we find Aries, ruled by Mars. We find the house itself to be empty, which means that Venus has no friends of her own. Only the ones she has stolen from Mars, the rightful owner of the true 11th.

Outside the 5th house, late in the 4th, we find Uranus. Well, Uranus, the planet of the quirky, "wants to be in the 5th," right? Well, no. Uranus is retrograde. It would rather not go there, it does not want to be "Venus' friend." With no rights to the 4th house cusp and not wanting to be in the 5th, Uranus is in limbo. In fact Venus has no proper friends (11th from the 7th, i.e., the 5th). With its ownership of Taurus and its indirect ownership of Cancer in the 8th, Venus — Wall Street — has bought its friends, as it seems that money is all it knows.

This confusion of 5th and 11th also accounts for the persistent talk of sexual assaults at OWS. Which, knowing a little bit about crowds, I doubt have occurred. (There have been no creditable reports.) Since the 11th of the chart is the 5th for Wall Street, Venus-Saturn-Libra make for kinky Wall Street sex. Which is widely believed to be true, even if it's not in the news. Like all good nags, Wall Street is projecting its habits — which it knows well — onto Occupy.

In this chart we see the simple bloody genius of astrology: It is often perfectly literal. A retrograde planet does not want to go forward, for example. Be literal. You will often find it to be true.

And here the first corner of the T-square has now been puzzled out. Venus and Uranus are exactly opposite, an aspect of battle, if not the violence we have seen. They hate each other. Venus is Wall Street. That Uranus is retrograde is why the protesters have not themselves been violent. Backwards motion inhibits Uranus.

Look now at the second house. In it we find Capricorn, intercepted. This is the money and resources of the protesters. We have already seen that the protesters' "obvious" resources (the 2nd cusp itself) to be ruled by Jupiter of the 6th house, represented by the police. Now we see that the protesters' fundamental resources to be owned by Saturn in Libra. Which we have identified as the government, and which we have shown to be "in bed" (perhaps literally) with Wall Street, represented by Venus in Libra, ruler of Saturn. Both planets quite content with each other. What do the protesters have? Nothing!

I confess I cannot remember when I have seen a chart this grim. And I am still not finished.

Astrologers typically read what they want to read and for this reason, if for no other, astrologers are taken as flakes and frauds. It would be easy enough to read this as a natal chart and say, well, Sun-Mercury-Venus in the 10th, the protesters will win, success is ours! Over at my Facebook page, where I developed this delineation over the past week, Laura posted a comment that a Republican pollster was allegedly "frightened" of the protesters. Which I have heard elsewhere.

Please snap out of this daydream and come into focus. Wall Street is "fearful" the way an abusive husband is "fearful."

Wall Street is the father who molests his daughters. Is he fearful? Is he a craven little coward? Who cares! The abusive husband, the molesting father must be stopped. Excuses change nothing. Wall Street (Venus in Libra) and the government (Saturn in Libra) are working hand in glove (the mayor of Oakland has said as much) and are like cats, playing with a mouse (the protesters).

Wall Street and the government have enormous resources and will bring as much of them to bear as necessary to stamp out the protest. For that is their goal. To date the struggle has lasted more than two months. It should be clear by now. The mice may be putting up an heroic struggle, but it is an entirely a one-sided and ineffective one.

If there are intercepted signs in the 2nd and 8th, the money houses, and if we interpret the intercepted signs as "savings" and the cusps themselves as "checking accounts," (astrology is literal!) then we have already seen what has happened to the protester's own money.

Now we look at the 8th house cusp. Gemini. A late degree on the cusp, Wall Street has already scooped up about as much loose cash as there is to be scooped. The sign being Gemini, Wall Street is a thieving swindler. We find the ruler, Mercury, to be in Virgo. Since Mercury rules Virgo, its deceit cannot be stopped, and, as Mercury is in the 10th and rules it as well, they are in full public view. Both Gemini and Virgo being mutable, and with a mutable planet, Mercury, in Virgo itself, Mercury will do anything, go anywhere, to keep on swindling and thieving. Gemini's immorality annoys Virgo, but Virgo will put up with it if the results are neat and tidy. All the i's dotted and t's crossed. Virgo is the fine print in Gemini's contract that takes all your money.

We find the Sun, the life of the chart, to be in the 10th, in Virgo. Owned by Mercury, the Sun gives his blessing to all of Mercury's many money-making schemes. While Wall Street's motivations (Moon in Taurus in Wall Street's 12th house) are well-hidden, nothing else is. Everything else is plain as day. No one can say later that they "didn't know."

We have arrived at the nodes. While they are technically in the 1st and 7th, their signs place them properly in the money houses, 2 and 8. The protesters get the north node, in Sagittarius. From what I read on-line (Firedoglake), as fast as the cops destroy protester property, donations from all over replace it. This is the nature of the north node, to pour in uncontrollably.

Wall Street gets the south node. Those Taurean cops (big, massive bulls they are, too) are expensive. Cities are strapped for funds and already we can hear that too much money has been spent. Given that the south node (a drain, sucking things down) is in Gemini, we may presume that light-fingered legerdemain to be manipulating accounts.

So now look at the T-square. Venus-Uranus opposed, squared by Pluto. All in cardinal signs. We have already seen that Venus, representing Wall Street, is opposed to and in conflict with Uranus, which is an unwilling (retrograde) stand-in for the protesters. What are they fighting about? (The fight is defined as the apex of the T-square.) They are fighting over Pluto, which, in the second house, represents the protester's (nonexistent) savings. Note that Saturn is the ruler of Capricorn, and that Mars is exalted there. While Saturn is not part of the T-square, his presence in Libra strengthens Venus' claim over the protester's property, just as the rights that Mars has in Capricorn strengthens the reluctant Uranus. Who will win?

One surprise in studying this chart was to learn that both traditional as well as modern rulers were "speaking." We find this with the third house, the media. With Aquarius on the cusp, the media has given a one-sided account, that of Saturn-Venus-Wall Street.

The "New Age" ruler of Aquarius is Uranus. We find it exactly opposite Venus, in other words, directly opposed to Wall Street. So we find the "new" media — the internet — to be behind the protesters and against Wall Street. We also find the new media to be, in general, opposed to the "old" media, as Uranus is widely opposed to Saturn. In the third house of media we find Neptune, retrograde. Neptune retrograde, the media, captured by the government and Wall Street, has been telling lies. (If it was direct, Neptune would be inspiring.) Judging by the late degree, Neptune has been telling lies for a long time.

By "long time," I do not mean "ever since Neptune went into Aquarius," as I am considering its house placement, *not* its sign placement. I mean that the media, represented by a retrograde Neptune late in Aquarius, has been lying for a very long time. Neptune merely identifies the problem. Neptune itself is not the problem. The media is. Like the big "X" on a treasure map. The "X" on a piece of paper is not where the treasure is. The "X" merely signals a location and enables you to find what you seek.

Pluto as "New Age" ruler of Scorpio, in the 2nd house of the protesters' savings, shows that we must have some new solution for money itself. Which I think we are, by now, all painfully aware.

Neptune as "New Age" ruler of Pisces (the 4th house of the occupied parks), shows, in its retrograde condition, the filth and squalor and drugs the mayors claimed was a problem and which the media (Neptune in the third) trumpeted. You will note the protesters themselves repudiated this rulership, claiming instead that they lost a great deal (Jupiter) of valuable property (Taurus). You might consider this as you use these two sets of rulers. I know that in future I will.

I come now to larger affairs. The First Amendment notably says "the people" have a right to assemble and petition the government. This is, in my opinion, a sloppily worded phrase, and for two reasons.

One, there is no requirement the government consult or recognize such assemblies. And, in fact, no government in the US has, so far as I am aware, ever paid the least attention to peaceable assembly. It is commonly believed that "protests" ended the Vietnam war, but some of us think an ongoing conscript mutiny was the real driving force behind the US pullout. By the early 1970's, privates shooting their commanding officers was becoming a common event. As Mao said, power flows from the barrel of a [conscript] gun. Which the Pentagon was learning, to its horror.

The reason protests are futile is because we are — allegedly — a government run by the people. If you don't like what the government is doing, well, vote. Vote 'em out of office. Vote in someone else. Nevermind that officeholders have every reason to manipulate elections get what they want, and have in fact done so. But voting is how you get change and "it's not the government's problem" that election after election never seems to change anything very much.

In putting this phrase in the First Amendment, the Founding Fathers had forgotten they were no longer ruled by an absolute monarchy. In a monarchy, when angry citizens pour into the streets, **the monarch listens**, as that's the only means the people have to make their voices heard. The monarch is monarch for life. He has his very life at stake.

By contrast, in elective forms of government, protests are ignored. By definition. We are your rulers today, our friends will be your rulers tomorrow, and their friends will rule the day after, so no matter what you do, you cannot get rid of us. Elections give the people the pretense of control. No form of government gives actual power to the governed.

Which Occupy Wall Street has made painfully clear. Protesters have been beaten and shot, pepper sprayed and tear gassed, arrested by the thousands, thrown into truly shocking and revolting prisons, their property seized and destroyed. In all of this, we have heard no murmur of protest from any elected official. Not from any mayor, not from any governor, not from the President himself. By contrast, we have seen him, and his Secretary of State, give vile lectures to other countries on the mistreatment of their protesters.

Instead we hear that unemployment benefits are to be cut, that payroll taxes are to be raised (the earlier cut was only temporary), that the age for Medicare is to be raised, that Social Security is bankrupt, that we are shortly to be required to buy worthless health insurance. For the last two years we have heard there might be a "field of dreams" infrastructure jobs project (if you build the roads, "they" will come), which has already been tried in Japan. Their now-excellent roads and rails are largely empty. But even this is talk.

It's not that Occupy Wall Street has lost. It's that they never had a chance.

Going forward, what comes next? Since charts "decay" (progress) over time, the OWS ascendant is three degrees away from decaying into Sagittarius. Chart ruler Mars is less than one degree away from entering Leo, and with it, the 9th house. Let's presume that decay is real and consider it.

Mars in Leo is a military man. The same decay that moves Mars into Leo also transfers it from the 8th, into the 9th house, just about when the ascendant becomes Sagittarius. While the 9th house can mean many things, its primary meaning, in America, is fundamentalist religion. Will we have a charismatic, born again Christian military leader in our future? David Petraeus, perhaps?

It's not simply that the two political parties have failed, but that they are unaware of the scale of their failure. We are desperate for leadership. Much as the Germans were in 1933, having been battered by WWI, hyperinflation, and economic depression, one after another.

This past week the Senate passed legislation giving the military (not the police) the right to declare American citizens, living in America, to be "terrorists" and to arrest and imprison them, indefinitely, without rights or process. It will be signed by the President and will be law by year's end. This has been widely denounced as the end of the Constitution. Which it is.

The rationalization for the new law, which I have heard, is that middle aged females are paranoid about monsters under their beds. Which is evidence the nation is polarizing even further. The upcoming presidential campaign is getting down to serious business at last, with the first primaries next month. I regret I find all the creatures running for the office, including the sitting president, to be unfit.

Welcome to 2012, the fated year. — *December 6, 2011*

🕐 *Occupy Part 2:* **Time, and Time Again**

Last week Thursday, December 8, a court in Boston declared Occupy protests to not qualify as free speech under the First Amendment. "Protesting" is free speech, but "occupying" is not, so said State Court Judge Frances McIntyre. Can you tell the difference? At sunrise on Saturday, Occupy Boston was promptly eliminated. Note this was the Saturn hour of the Saturn day, Saturn representing government.

Ignore the happy talk, if you can still find any. The Occupy movement has been suppressed by the police, who are clearly acting in concert with Homeland Security, which is under the direct supervision of the President himself. Truman told us long ago where the buck stopped, Nixon confirmed it, Bush removed the last doubts. Not only is the President The Decider, his office is, as a whole, charged with the Manufacture of Reality itself. Presidents give us their permission to react. Obama is no different.

Over the past week various people have suggested other "birth times" for Occupy, which officially started on September 17, 2011, 12:01 pm in lower Manhattan. From Anne O, repeating an observation from an astrologer on the scene, 10:01 am, the moment when Occupy physically sealed off the greater Wall Street area itself. Gary P suggested 3:00 pm, when the General Assembly first met. Gary also suggested July 13, 2011, as the start of Occupy. That was when the September 17 date was first proposed by Occupy's original promoters, the Canadian based Adbusters.

All of which produce interesting details. 10:01 am shows underlying public support, which is considerable. A youthful, powerful, sexy 3 degrees of Scorpio rises. It owned and controlled by the same worn out, exhausted, nasty Mars, but that Mars is now in the very center of the 9th house, of religious beliefs, hopes and dreams (dreams are 9th, not 12th). Mars is still ruled by the Moon, but the Moon is now late in the 7th. A retrograde Jupiter stands between it and the descendant. And while Venus in Libra still rules all three, Venus is now not in the 11th, but rushing towards the 12th. Venus is not as strong, Jupiter is standing in the way, which means the Moon is weaker overall. Which makes Mars stronger as a result, but, alas, not a lot stronger.

As for the Wall Street crowd, 10:01 am gives them 3 of Taurus on the descendant. If the crowd's enthusiasm (3 Scorpio) was, in fact, premature by two hours, then note that Wall Street itself, at the equally early degree of 3 Taurus, was "caught off guard" by the events. As was its ruler, Venus. Venus, at 3 Libra, was exactly inconjunct to the descendant and thought she was safe, that "nothing could happen" at such a time. She was to learn otherwise.

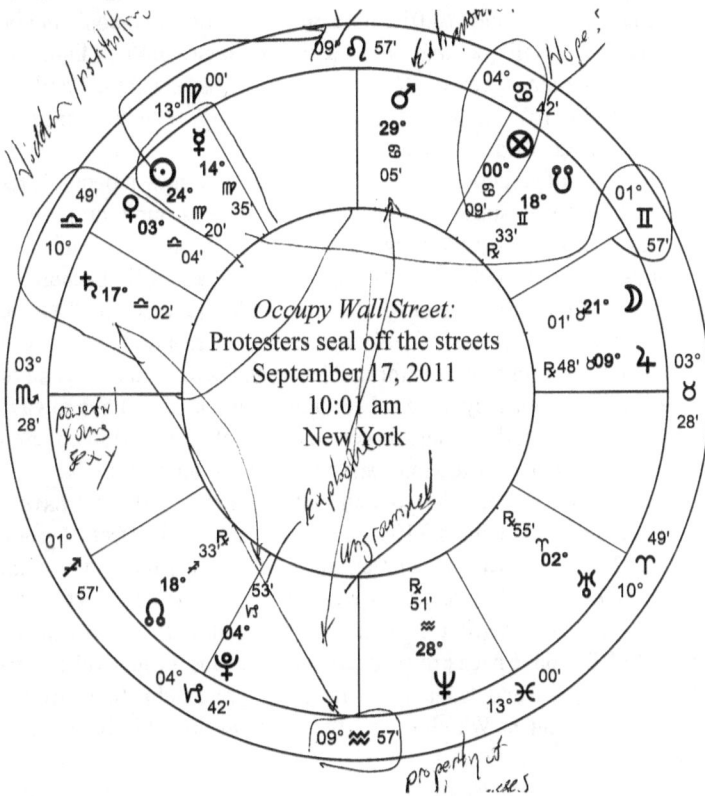

Venus ruling 12 and going in that direction, Wall Street immediately sought institutional (12th house) help. Where it found the government (Saturn in Libra in 12) ready and eager. Note that at 10:01 am Saturn rules the 4th house, which represents the physical space Occupy was occupying. Note that the "new" ruler of the 4th, Uranus, retrograde, was trying not to get into the 6th house. The 6th is the police, Aries on the cusp says the cops would be aggressive, but **Mars ruling the 6th says**

the police were on the side of the protesters. Which in New York they initially were. Police suppression in New York was originally from white-shirted supervisors. Not the blue-uniformed rank and file who stood to one side, watching. Police hesitation is shown by Uranus in the same sign as the cusp, but located outside the house and going backwards. The police wanted nothing to do with arresting protesters.

Which is evidence of the "fog of war." The 10:01 am chart shows a popular uprising. Uranus in 6 opposite Venus and Saturn in 12. The need for police (6th) to repress the revolt was clear, but the police, in contact with events on the ground (10:01 am, loyal to the protesters) could not be relied upon. You will note that both Venus and Saturn are debilitated in Aries. Which means "they can't go there." Wall Street (Venus) and the government (Saturn) were both frightened, being opposite the explosive Uranus, in the sign of their mutual debility.

The story of Occupy then becomes the story of how a people's revolt (10:01 am) was subsequently co-opted by the group's promoters, Adbusters (12:01 pm, see chart, pg. 3). Searching on-line, I can only find that Adbusters was founded at some point in 1989, allegedly because of rage at a vending machine. Note that the Berlin Wall effectively "fell" at 10:30 pm on November 9, 1989. If Adbusters was a reaction to that (as it likely was), then Adbusters was formed between that date, and the end of the year. Like as not they have a 12th house Sun, as they have always been institutional, shadowy, and largely ineffectual. Carol Mull covered 12th house Suns decades ago in relation to corporation charts.

Occupy failed when the government realized they did not have a popular uprising on their hands (10:01 am), but merely an ineffectual group of fools (12:01 pm). The government then acted accordingly, and with great strength. If that is true, then we might get more information by setting the chart for the first General Assembly, at 3:00 pm. Evidence suggests OWS to have been pre-planned, which means the general assembly was pre-planned and its inaugural session to presumably to have been scheduled in advance. Which means the 3:00 pm time will be more or less accurate.

Three pm on the fated day has 5 Capricorn on the ascendant. Pluto is exactly conjunct. Capricorn is structure. Pluto wants to destroy that structure. As with the previous two charts, we are no longer surprised to find the chart ruler again in the 9th house. The chart ruler, the speaker of the assembly, is Saturn, which shows authority. This authority is (take your pick) a philosopher, a clergyman or a foreigner, as the 9th rules all three.

These General Assemblies were immediately suppressed. Use of amplification was prohibited, forcing a nuclear-powered Pluto, ruled by a powerful Saturn booming from the ivory towers of the 9th (both of these planets malefics), to a slow, unamplified "call and response" method of deliberation.

If the ascendant is the speaker, the seventh is to whom he is speaking. Here we find Cancer, with Mars at the very last degree. Mars is wrapped up inside himself at the best of times (a man with a sword isn't taking direction from anybody!), and at 29 degrees (back to the wall) is certainly in no mood to listen.

Which makes the 10th house the subject under discussion. At 29 Libra, the subject, whatever it is, is stale. Venus ruling from the 9th, which is the 12th from the MC, tells us the subject is hidden. The final outcome is the 4th house. At 29 Aries, with Mars exactly square, the affair ends badly as well as abruptly.

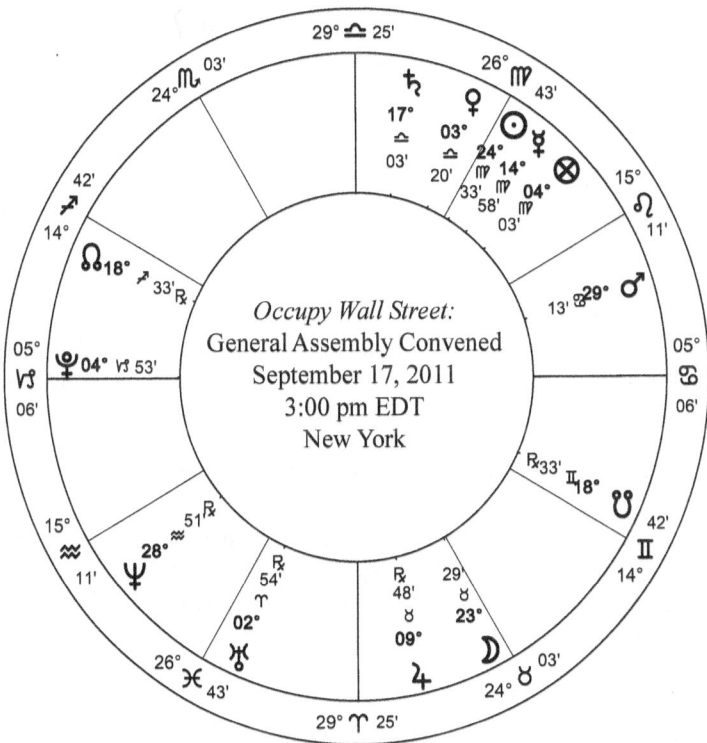

Which brings me to the chart for this country, the United States of America, July 4, 1776, Philadelphia. The original chart, an artificial construction, was first published by Ebenezer Sibley in 1787. It consisted of the angles and houses from the previous summer solstice, as seen in London, combined with planetary positions as of noon, Philadelphia, on July 4, 1776.

Thinking about this, two things are clear. One, Sibley, like many ancient astrologers, did not have the actual time of the event and so used a workaround. When you study ancient texts, watch out for this. A lot of old techniques would never have been invented had they been in possession of accurately timed charts and exact ephemerides.

A revolutionary declaration is, logically, a subset of the prevailing solar cycle, which makes the most recent phase of that cycle, the solstice, a valid starting point. The second point is that Sibley most likely saw the American revolution as a temporary arrangement. Not a permanent condition. A more permanent chart would have been used the angles from the spring equinox. Which, then as now, is the proper start to the year.

In the *Book of World Horoscopes,* where I extracted this information, Campion notes that Sibley published his chart only eleven years after the Declaration, only five years after the war ended, and, revealingly, two years before Washington became the first president. Which *"was therefore calculated while the events of 1776 were very much a part of contemporary politics."* (pg. 414, 1999 edition)

The importance of this seems to have been lost on Nicholas. Which is that five years after the war ended, the thirteen colonies still had no effective government. The likelihood was that the states would sink into petty warfare and if one or another state did not end up the victor over the rest of them, the colonies would eventually be reabsorbed by one European power or another. I would like very much to see Sibley's original delineation. I am also curious if he set another chart upon learning of the Constitution and the inauguration of Washington as the first president. If you're curious, Campion gives that as April 30, 1789, 12:40 pm, New York.

The Declaration of Independence was forced by the king of France, the luckless Louis XVI. The rebels had gone to Versailles to ask for help. The French were sympathetic, having lost a lot of real estate to the British a decade before (Quebec), but rightly insisted the colonists make a formal declaration of their intentions. It would seem, from the rather murky accounts of July 4, that they did so reluctantly. We might further speculate they did so as a group so that the governors of the individual states could deny the document later, if needed.

The resulting astrological chart, whatever time you set it for, seems to have never actually been read. Some degree of Sagittarius or maybe Gemini (the ascendant) is known to be a significant degree, and that's about it. So let's read the rest, see what it tells us:

Occupy Part 2: Time, and Time Again   15

At 4:50 or 5:10 pm, times which seem to have historical support, Sagittarius rises. This is enthusiastic and outgoing and energetic, but also mindless. The ruler, Jupiter in Cancer, is technically late in the seventh house, but as Gemini, not Cancer, is on the cusp of the seventh, it seems to me that we should put Jupiter, along with the Sun (which wavers between the two houses between the two times) and Venus, in the 8th, where it will join Mercury and the Part of Fortune.

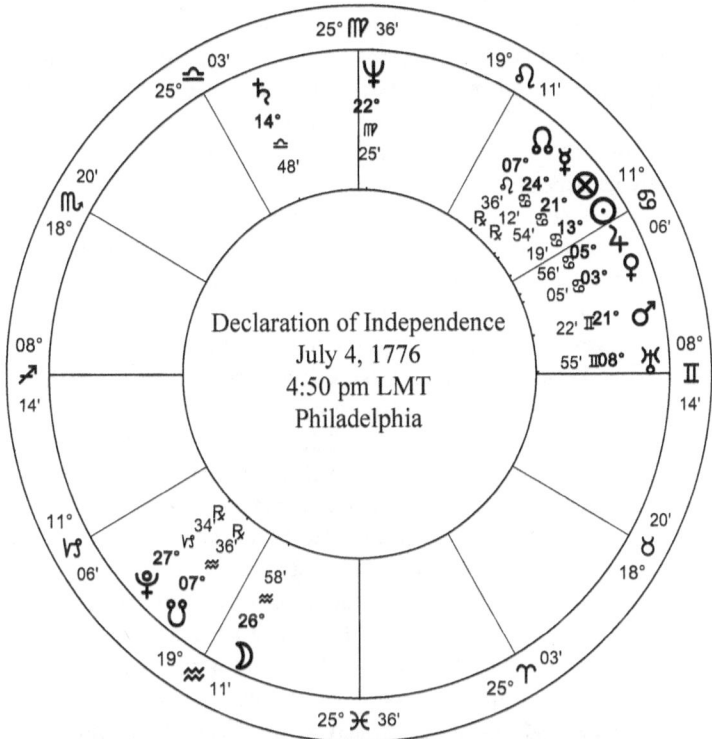

This gives us a stellium of four planets and a strong 8th house chart. Stelliums are strong because the planets in them will rule, and therefore dominate, much of the rest of the chart. In this case:

Venus in Cancer and the 8th rules the MC (5:10 pm), the 11th house, and the 6th.

Jupiter in Cancer and the 8th rules the ascendant and, at 4:50 pm, the fourth, Pisces.

The Sun rules the 9th.

Mercury rules the descendant and the MC if the time is 4:50 pm.

As no planet is in its own sign (Jupiter is exalted), every planet not in the stellium is ultimately ruled by the stellium itself. The Moon is the ruler of Cancer. We find it in Aquarius in the 3rd. It is ruled by Saturn in Libra, which is ruled by Venus in Cancer.

So our 8th house stellium can reach out and claim the entire chart for itself. Directly with the ascendant, 4th house, 6th house, descendant, 8th house, 9th house, 10th house, and 11th house. Directly, the stellium rules the north node, the Part of Fortune, Neptune and Saturn. Indirectly (one stage removed) the stellium rules all the other planets (the Moon) and all the other cusps. As we have seen, the 8th house is the house of Wall Street, the house of other people's money. With a stellium in the 8th, one which includes an exalted chart ruler, this is a chart of a nation which will tend to exploit itself as well as the rest of the world, for its own, personal benefit. (Cancer takes everything personally. In the 8th, it helps itself at your expense.)

You will note that Venus, a part of the stellium, directly rules the 6th house, which has Taurus on the cusp. Which in traditional astrology was the house of slaves. Taurus is money, which is to say there is money to be made from slaves. That Taurus is empty denotes the relative unimportance of slaves, as well as the unimportance of their successors, modern wage slaves. Which, in a single stroke, demolishes the fiction that America was founded to be free. Slaves and slavery were an accepted part of the arrangement. When we read the Declaration of Independence we read much about The King did this, or The King did that and as a result we believe The King to have grievously oppressed his people. We ignore the fact that half the signers were slave owners (no, slave *breeders),* and we are conveniently ignorant of what that means.

In fact, by the mid 1770's, the abolition movement was well underway in London, led by the tireless William Wilberforce, a name little known in the US. In an effort to make the rich northern colonies make peace, England proposed abolishing slavery in the colonies. After all, the northern colonies already had by that time, so slavery was clearly on its way out, no?

Instead, the offer was rejected by Southerners, who then had a reason to join their northern cousins in revolt. Hence the reason why the Declaration did not list slavery as contravening the Rights of Man, because so far as America was concerned, slavery did not. England banned slavery in 1807, and as the English ruled the seas, they banned the Atlantic slave trade as well. The US government nominally went along with the trading ban, but in fact kept on with slave ships. Which the English navy seized, impressing their sailors into the Royal Navy for good measure. Which set up the War of 1812. They didn't teach you this in school?

Cut off from cheap resupply, hereditary slavery requires slaves to be sired by their owners, as hereditary slaves will not breed. ("Let slavery

Occupy Part 2: Time, and Time Again   17

die with me" is every slave's motto. Slave mothers who cannot abort will strangle their newborn, with tears in their eyes.) In the 1776 chart, count five houses from the house of slaves, to find the children of slaves, and note the ruler (the de facto fathers), and its placement. I had not looked at this chart detail before but was not surprised to find it. I could say much more, but it will turn your stomach. Slavery in America was far more vile that the fairy tales we have been told.

Note that Uranus, the planet of freedom, is in the 6th at 5:10 pm, but in the 7th at 4:50. With the change of Virgo for Libra on the midheaven, I am coming to the conclusion that 4:50 pm is the more likely time.

The Sagittarius rising chart exists to extract money from the world. It is interested in very little else. Those who oppose it will be struck down: Mars in the 7th. Americans have traditionally "found themselves" inside this chart, which is to say, in its 8th house and have consequently enjoyed themselves at the expense of others: The original inhabitants were pushed aside, African were forcibly brought to do the heavy work, and immigrants have traditionally been exploited ruthlessly. Chinese immigrants have horrible stories to tell, to say nothing of the present migrant worker mess. Until quite recently America consumed most of the world's resources and still consumes disproportionate amounts of oil and energy.

America has always lived at the expense of others. We are an 8th house nation.

The 8th house is the house of transformation, the house that takes things, such as money, from one place (your pocket) to another (its pocket), or, on a personal level, transforms you from virgin to sexually accomplished, or takes you from life, to death. These 8th house processes are innate. They cannot be stopped, they can only barely be understood.

Do you remember, years and years ago, what it was like to be a virgin? How you wondered, how you looked forward — or dreaded—the day and the place and the person that would change your life, once and forever? *The process of losing one's virginity is ruled by the 8th house.* Death, of course, is the primary 8th house function and a much more powerful and frightening example.

The 8th house, the house of transformation, is the essence of the United States of America, now and forever. Regardless of prudence or cost. Eventually a chart like this will turn on itself.

Now do you understand what America is? It's in the chart, for all to see.

Do you understand, now, finally, just how high the stakes are, what is on the table, and what Occupy was really being asked to ante up? What

Wall Street and the government so deeply feared? Does the 1776 chart still work? Is it still powerful? Truthfully, no one knew, as it had not been tested in many years.

Now we know. In the 1776 chart, the ruler of the 6th, the house of slaves, is in the 8th. This empty 6th house means the slaves and workers of America, the 99%, have no voice. The 1% of the US, the financiers, the bankers, the slave owners, have a long history of repressing slave rebellions as well as labor unrest (Shays' Rebellion, 1786-7, etc.). The fate of Occupy Wall Street was preordained.

One could imagine a revolution, a new Constitution, new leaders, to sweep all before them, resulting in a "new birth of freedom," to use the old phrase. Is this possible?

The outcome of revolutions can be found, again and again, in the pages of Nicholas Campion's masterful *Book of World Horoscopes*. Campion gives charts for every French Republic. Gives charts for every flavor of Russian revolution, as well as numerous post-Soviet charts. Study all of them, and I learn that **the underlying nature of a nation does not change**, from constitution to constitution. America is an 8th house land and always will be.

It could be that, just as Ptolemy divided the nations of the world among the twelve signs of the zodiac, that nations can also be divided among the twelve houses, and that America is the principal 8th house country in the world at this time.

Imagine that once upon a time, many millennia ago, America was part of the world, and imagine America was even then an 8th house place. And imagine that America misbehaved and as a result was cut off from the rest of the planet and the continent left to "rest" until perhaps its overwhelming intensity (shall we say) had abated a bit. What do we do, we who must somehow live here? I regret I have not a clue. — *December 13, 2011*

Postscript, July 2012: There was a recent report that Adbusters was a government front group, designed to organize, and then discredit, American protesters. While there are always stories like this and while one should take them with a grain of salt, it is true that Adbusters is in fact a mysterious group, that, aside from the 99% slogan, Adbusters was useless as a leader or organizer, and that the various governments, city, state and federal, individually as well as a whole, immediately and with a great resolve, set out to suppress the movement. Astrology clearly shows the movement to be inept. We do not know the reason why this was so.

☉ *The chart of the week:* **Howie Mandel**

When I reluctantly signed up for Facebook a month ago, I had a quandary. On the one hand, I didn't feel I had the right to ask strangers to "be my friend." On the other, I have become an astrological heavyweight. A lot of the other astrological heavyweights have Facebook pages and if I didn't "friend" them, they would think I was stuck up. When it's really too much Saturn that makes me shy.

So I decided to friend all the biggies who turned up in the lists, despite the fact that many of them have written books that I've not been kind to in my reviews. Which, by the way, is from too much Moon/Pluto.

One such heavyweight was Glenn Perry. I asked him to be my friend, and a day or two later, he accepted. He then asked if I would sign up for his newsletter, so I did. Professional curiosity made me go to his site to find out what kind of newsletters he was writing, but I found he doesn't post them on-line, only the columns from them.

Among Glenn's columns, I found one on Howie Mandel. Glenn likes Howie. He says Howie has mental disorders, and when I looked into it, I found that Howie himself says he is disordered: Obsessive Compulsive, Attention Deficit Hyperactivity, as well as Manic and Mysophobia, which is a pathological fear of germs.

With this as a platform, Perry proceeds to delineate Mandel's chart. My personal opinion is that psychological astrologers would be greatly helped if they knew a little real astrology. So let's read Howie's chart and help Glenn out.

Howie Mandel was born on November 29, 1955, in Toronto, Ontario. The 10:00 am birth time sounds approximate and we may tweak it as the analysis proceeds.

Howie Mandel was born on a full moon as well as *during a lunar eclipse.* This is an astonishingly rare birth and by itself sets Howie apart from just about everyone else. How do we know Howie was born during an eclipse? We see the Sun at 6 Sagittarius, we see the north node at 17

Sag, and the Moon at 5 Gemini, and we automatically know there was a lunar eclipse of some sort. **This should be hard-wired in your head**, that when Sun and Moon are conjunct, or opposed, and the nodes are very nearby, **you have an eclipse event**.

So we go to the ephemeris and we find that, indeed, there was a partial lunar eclipse that peaked at 17:00 GMT. Which was noon, EST, in Toronto. Mandel was two hours old.

Knowing that lunar eclipses are hours long events, we dig into Google to discover the eclipse started at 9:51 am EST, nine minutes before Howie Mandel's stated birth time, when the Moon first grazed the Earth's penumbra, the outer limb of its shadow. This particular eclipse was visible in Asia, but as it only barely touched the umbra (darkest area of the Earth's shadow), even at its peak, Mandel's eclipse didn't look like much. The Moon appeared to be a bit dimmer, and that was it. So, having uncovered a rather weak eclipse, we are wondering if, in fact, it had much impact of its own. As we do not know, we set this part of Mandel's chart aside. We will come back to it should we find others born during a lunar eclipse, as we need more than one sample to make a judgment.

Howie has an 11th house/5th house Full Moon. Where have we seen something like this before? It was the chart of former House Speaker and current presidential candidate Newt Gingrich, which I featured July 26, 2011. (See *Duels At Dawn*, pg. 112.) In Newt's chart, the Sun is in Gemini in 5, the Moon is in Sag in 11. In Newt's case, the Sun runs around with females (5th), while the Moon seeks approval from friends (11th).

Mandel has the same signs associated with the same houses (note the cusps are the proceeding signs), but the lights are reversed. Why do I not consider Mandel's Sun and Moon to be "running to get into" the 12th and 6th? Because in his life Mandel has shown no interest in institutions, nor medicine per se, which would be a 12/6 axis. I am in agreement with Glenn Perry, that Mandel's interest in germs comes from Jupiter in the 8th as the apex of a mutable T-square, but I have a better explanation for it.

Like Gingrich, Mandel's Full Moon is fascinated by an ever-changing reality that he cannot control, only observe in wonder. Unlike Gingrich, the mental component is missing. Mandel's Mercury is debilitated in Sagittarius. It eagerly takes up with every new craze, with each new fad. As we have seen with other charts, Mercury debilitated wants to migrate to the opposite house, the 5th, where it finds the Moon in Gemini.

The Moon in Gemini is a scatterbrain. It jumps nervously from idea to idea, it does not know what it is doing, its emotions are light and superficial and of no consequence. Moon in Gemini opposite (dead opposite) Sun and Mercury, both in Sagittarius, this is someone who can suffer from extreme nervous stress. From five to eleven, the individual will constantly invent (Moon in Gemini in 5) events to impress the friends he wants to have, but which, in fact, he does not.

Follow this carefully: Sun and Mercury in Sagittarius in the 11th is the need to have friends, lots of friends, but Capricorn on the ascendant is a sign of unpopularity (too severe, too "old"). Chart ruler Saturn in Scorpio is just outside the 11th house and really does want to "run to get into it" and take it over. Where his negative (Saturn) intensity (Scorpio) overwhelms and drives friends (11th) away. See what you can do with simple keywords?

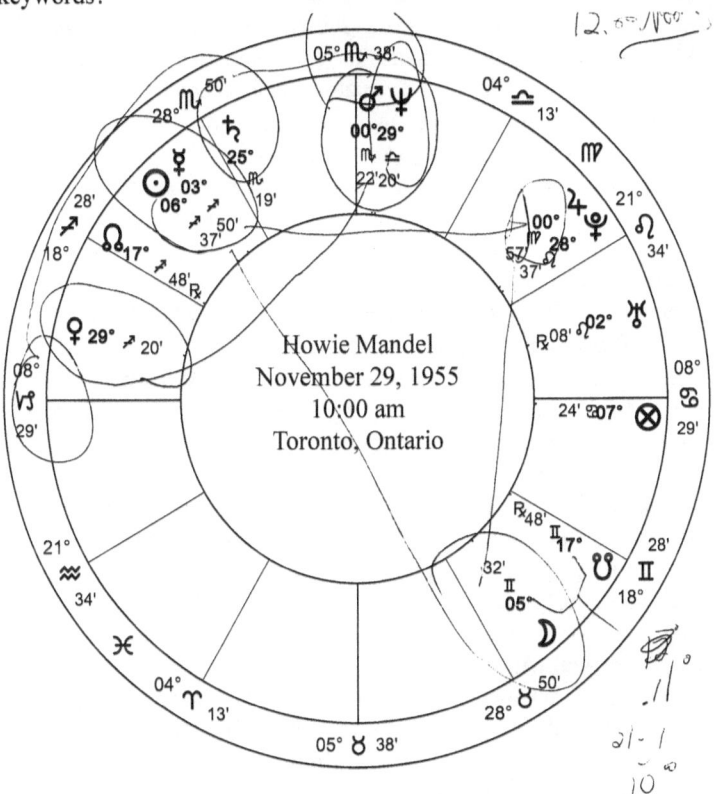

As a result, Mandel's Sun-Moon opposition is constantly thinking up ways to impress people. To get attention. To have friends. Full Moons are intensely dynamic, they possess infinite range, they can carry on beyond imagination. (Trust me: I have one myself.) So, on the one hand, chart ruler Saturn is taking Howie's friends away, on the other, Sun-Mercury opposite Moon from 11 to 5 is a spectacular show-off. If Howie is seeking attention, the lights are powerful enough, in their own right, to go on forever. With Gemini and Sagittarius as the signs, the result can be complete nervous exhaustion.

As a full moon birth I suffer from nervous exhaustion, and while my Sun and Moon are in fixed signs, not mutable ones, they are in cadent houses normally associated with Gemini (3) and Sagittarius (9). So I am wondering if Full Moon Howie is more of a nervous wreck than I am. In addition, Mercury, planet of nerves, is part of my Full Moon configuration, just as it is with Howie.

Howie Mandel is so nervous he is on medication. The underlying problem is not that he is nervous, but that he is showing off to impress friends, and will compulsively go on showing off until he is accepted by them. Maybe if someone explains this to Howie, maybe if he gets a personal People's Choice Award he can at last feel accepted and stop acting out, stop being hyper and so stop taking drugs, as drugs are evil. (He was part of St. Elsewhere when they got a People's.)

Instead, Howie has become the spokesperson for those with Attention Deficit Disorder. This is unfortunate because labeling oneself as ill strips one of the power to heal. Instead, one becomes passive (powerless) and runs to others to cure him, or becomes anal and uses his condition to manipulate. Both outcomes are undesirable. I considered that perhaps ADHD sufferers were Full Moon people, but that is unlikely. There are planetary positions that produce powerlessness, but a full moon is not one of them. A full moon is the most dynamically empowering of all aspects, but it does require the native to fully live it. I know this well.

A couple of days ago I sent a friend an email where I explained T-squares as an opposition that is "fighting over" the apex, which is to say, the planet in square to the opposition planets. Two men fighting over a girl, for example, or maybe three in a bed, which is always two of one sex (the opposition) and one of the other. Which, if you take the time to look at a T-square, you will find to be true. A T-square is nothing more than a love triangle. With Howie Mandel, I want to modify that a bit. In Howie's case, Sun-Mercury opposite Moon are driving Jupiter, square to all of them, to extremes. They are beating up Jupiter. They are bullying him.

They can do this because Jupiter is in detriment. If he were not in detriment, he could fight back. He could establish himself in Virgo, which at 10:00 am (from memory) is intercepted in the 8th. Intercepted means Jupiter has no cusp, no "anchor," to call his own. Yes, debilitated Jupiter would much rather be in Pisces on the opposite side of the chart, but there is no anchor for him there, either. So he is stuck, weak and helpless.

In his delineation, Glenn Perry notes, correctly, that Virgo is concerned with cleanliness, but Perry is not well enough read to have found

Alan Oken's remark that the 8th house concerns germs, among other things. Virgo in the 8th, there is a nightmare of germs which a weak, disabled Jupiter is unable to get away from, which he magnifies out of all proportion. Jupiter's keyword is to magnify. He will magnify the bad as well as the good.

What are Mandel's true feelings? Jupiter in Virgo in the 8th without moorings, Mandel is terrified he will die from germs. "Without moorings" means Mandel has no control over this belief. (Which, by the way, is one of the ways a natal chart can show powerlessness.) This is a very real fear and one that is very likely based on his most recent death (past life). It must be respected.

Which, off the top of my head, sounds like a death from a virulent plague. Water-based, germ-borne plages were the curse of the Piscean age, which ended in 1913. During the Piscean age, with its emphasis on water, there were occasional great plagues that killed millions over entire continents. There were also many "minor" plagues that broke out locally every few years. Plagues of one sort or another lasted into the early 1900's. Once air-based Aquarius arrived, waterborne plagues disappeared. We now have mass virus attacks. Viruses are airborne. The first big virus? The "Spanish flu" of 1919. The Ages are real.

Unlike viruses, plagues were extremely rapid, because water is a far more deadly means of transmission than air. During times of plague, victims would awaken in perfect health but be dead by nightfall. No virus is that fast. In times of plague, one lived day by day, never knowing if today would be the last, praying every minute that the plague would end before it killed you, before it killed your loved ones. Sheer terror. (Read between the lines and you will find a first-hand account in Book 3 of William Lilly's *Christian Astrology*.) One who dies at such a horrible time will reincarnate with a justified terror of germs. Medieval doctors had the best understanding. Richard Saunders, passing along Dr. Forman's account, attributed the London plague of 1593 to fresh herrings and cucumbers, and while such a combination might produce cases of food poisoning, a plague is something far beyond that. But I digress.

Now watch how the rulers play out: Mandel's debilitated Jupiter in Virgo (not clean enough) is disposed by a debilitated Mercury in Sagittarius, which is to say, a weak Jupiter is infected by something bad brought (Mercury, messenger, debilitated) from far away (Sag). Something which Jupiter itself wanted, as not only are both planets debilitated, they are in mutual reception. Each wants what the other has. Now look at the cusp of the 8th. It's Leo, the sign of singularity. Just inside the house itself, we find the planet of deadly intensity, Pluto. Both Pluto and the 8th house

cusp (the house of death, remember) are ruled by the Sun in Sagittarius, which is the sign of foreigners and the *ultimate* source of the disease. That the Sun is also in the 11th shows the *immediate* source to have been among Mandel's friends. If you believe, if you know in your bones that in your last life you, in fact, died of an infectious disease you got from touching others, then in your next life you will go to any length to avoid human contact. It will be a matter of life and death.

You will further note that Mercury rules the fingers, and that fingers are also ruled by the sign Gemini, where we find the Moon, which represents the physical body. Fingers are what we use to touch one another. Is it a "phobia" when one has personal experience, past-life memories? If you were susceptible in your past life and in fact died as a result, can you ever be certain that you are not still susceptible, that you will not die all over again, from the same cause?

These simple things, a sign, a planet, a rulership, when properly read, can produce details of events of long ago. These details make a chart, an individual, come vividly to life.

A lot of people have simple ideas about past lives, that if you died of something in one life that you are thereafter finished with it and it will not bother you again. As I am privy to a many details from my own most immediate past life (a minor historical character with an actual biography), I find I suffered from arthritis in that life, and suffer from arthritis in this one. I had an undefined lip problem in that life, and an eternally split upper lip for most of this one. I was blind in my right eye in that life and have a faint birth scar on my right temple in this life. Not only can a tendency to specific diseases and health problems carry over from one life to the next, but unfavorable events in one life (certain forms of violent abuse) can create problems in the next (epilepsy, which I do not suffer from). How much more creepy can it get?

There are many who believe that past lives are shown in the chart by the nodes, perhaps with a boost from Saturn. I have worked extensively with past lives and the natal chart. In my opinion the entire chart can be read as a reflection of the individual's state as of the end of his last life.

Howie Mandel is a good example. It is not unusual that a planet's sign indicates one thing, its house placement another, as with Mandel's Sun.

So as Howie gets older and weaker from age, I would expect Jupiter to become ever more dysfunctional and for Mandel to become ever more germ-phobic. Intense phobias that are otherwise unexplained often have a grounding in traumas from the most recent past life, often the death. You need only take the time to work out the details, and then remember your results will be provisional. Never anything that can be known with certainty. So be modest.

Aside from the intensely nervous Sun-Moon opposition, none of this strikes me as "mental illness" material. Astrology is the science which eliminates the unknown, which conquers fear, but only if we dare to use it.

Now we put Howie's pieces together. His desperate need to impress people and have friends is flatly contradicted by an ever more extreme phobia about actually touching them. These two tendencies cannot exist in one chart. You cannot want friends but then be terrified of them. One trait must win over the other. I briefly considered past life leprosy as the connecting factor, but lepers commonly live with other lepers and are thus not entirely without friends. Nor are lepers prone to intense germ-phobia.

We can see Mandel's desire for friends/fear of friends symbolized by the not quite conjunction of his Mars and Neptune, in Scorpio and Libra, late in the 9th. If Mars and Neptune were in the same sign, we would get the classic "sick energy" or "sick action." Instead, Mars has passed over Neptune and has landed in his own "home" sign, where he breathes a sigh of relief. Yes, Neptune is still very nearby, but now, with the "door closed" behind him and Neptune on the other side of it, Mars can come to his senses again.

For those of you still having problems with "smeared" aspects such as this, consider my theory that the signs of the Zodiac are internal Earth energies, and then consider these energies to be similar to the instruments in the orchestra. A clarinet ("Libra") and oboe ("Scorpio") may play the same note and even be made of the same material (wood), but the sounds they make are not the same. The Earth's zodiac is made of twelve different frequencies, twelve fundamental "notes," each unique.

In Mandel's chart, Mars rules the midheaven, where he "stands behind" his public persona, ready to come to his aid if needed. As ruler of the 11th, Mars also compounds Saturn's baleful influence. Get too close and Mars will attack. This once happened with a man who mistakenly tried to shake Mandel's hand. Mandel took it as an assault and reacted accordingly. In other words, for a brief instant Mars leaped out of the 9th house and into the 10th. Mars owns it, after all.

So, as a whole, what do we have with Howie Mandel? We have a man who has a hard time making friends because of an unfriendly ascendant. Chart ruler Saturn's house placement compounds the problem, denying friends altogether. If the 11th house were otherwise unoccupied, Mandel would largely accept this. But instead he has set about to solve this problem, and by the most powerful means possible.

Buried in the 11th house we find Sagittarius, the sign of freedom. We also find the life-giver itself, the Sun. In a nervy, daring exercise of pure power, the Sun has put the Moon directly opposite, where the two of them can bounce things back and forth.

I am of the opinion that a Full Moon is dual. On the one hand it shows ultimate conflict, on the other the absolute desire to burst through all barriers and solve all problems. It is the second quality which is most important. Full moon births, especially tight ones, like Mandel's, are not accidents. I believe tight full moons are evidence of severe past life distress and an overwhelming desire to come to grips with the underlying problem. Consequently Full Moon natives have an enormous intensity.

In Howie Mandel's case, his primary problem will be resolved by means of friendship and acceptance. This might not be where the problem started, but it is where it has ended up.

Howie's secondary problem, death from plague, must be fully processed and finally, in the end, accepted and released. By comparison to the friendship problem, the fear of germs is trivial. I have heard of hypnotic regression solving such things. In Mandel's case, lack of proper diagnosis has compounded his fear of germs.

Taken together, I would expect these problems to require a lifetime's work. This, in the end, is why we take birth, to work out specific problems. Which will most likely require an entire life. Life is not a trivial thing. It is never entered into casually, you will be glad for every second that is given to you.

In his delineation of Mandel's chart, which if you print it, runs ten pages, Glenn Perry says, "*It would be significant, therefore, if it could be shown that mental disorders have clear astrological correlates that precede presumed biological or environmental causes. This should not imply that horoscopes cause psychopathology . . .*"

There is a certain mournfulness to this, as if Perry does not expect an answer. The term "mental disorders" is a misnomer. Astrological analysis is far more specific than vague catchall terms. Nor do "horoscopes cause psychopathology." No such thing is possible. The horoscope shows how the individual has reacted to his situation. The horoscope shows what the problems are, and what the consequences will be. Properly read, a natal chart is an infallible tool. Since the natal chart subsumes both biology and environment, it is superior to both. In the strictest sense, a natal chart is a mere *dossier*.

Healing starts with the understanding that we are not watching an isolated actor, alone on a darkened stage. Instead we have an interlocking series of events, an endless stream of players, some of whom have fared better than others, all of whom need our love and support, for without every one of them, we ourselves cannot live, as we are one among them.

I hope these remarks are encouraging to psychological astrologers.
— *December 20, 2011*

◷ Losing Your Virginity

Since all the readers of this newsletter have already lost their virginity, this will be along the lines of a postmortem.

Since astrology describes the energy of the Earth in its ceaseless relationship with the Sun, Moon and planets, it logically follows that astrology defines and describes everything that happens, everything that can happen, on this planet. All we need to do is find the formula and then apply it.

Unique among the twelve houses, the 8th house rules transformation, in other words, dynamic processes in general. (For the moment we are not interested in the other things ruled by the 8th house.) In addition to death, the 8th house has two distinct rulerships over sex. The 8th house rules married sex, in other words, the static duties the husband has to the wife, and the wife to the husband. And it rules virginity, or more precisely, the dynamics of its loss.

The setup will be the same as for death. A transit or progression to the 8th house cusp, or its ruler, or planet in the 8th, will time the event. The nature of the event will be shown by the 4th house, which traditionally shows the End of Things, as in, the end of your virginity. Which will be expressed by the nature of the sign on the cusp of the 4th, its ruler, and planets in the 4th.

At first I thought that was it, but as soon as I read for myself, I realized the 4th house was merely the stage and the actors were the 5th house, the cusp, its ruler, and planets in the 5th.

You are all grabbing your charts and seeing how it played out in your life, many, many, many, many years ago. (My how old we all are!)

One woman with Pisces on the 4th house cusp was so drunk (Pisces: unfocused water) that neither she nor her boyfriend were quite sure they had actually done it. So, south node in Cancer on her 8th, the next night they went back to her house (Cancer) stone sober and did it for real.

Another woman had Gemini on the 4th house cusp. Its ruler, Mercury, was conjunct Saturn in the 7th in Virgo. The 7th house being the house of the partner himself, her first lover was both quick witted and older. From the day they first met, until they first did the deed (they were both virgins), nine months passed, which is also Saturn and I rather think, Virgo as well.

In my case, I lost my virginity when my 8th house cusp had progressed, by solar arc, to become the solar arc descendant. In other words, the 8th house "ran into" the partner. The 4th house being Leo, the woman I was with was "singular", in other words, I never met any of her friends. The ruling Sun in the 9th house, she was a stranger to me. Virgo on the 5th house cusp, the event was intellectual, rather than emotional. Saturn intercepted inside the 5th house, the event was incomplete, and, Saturn striking yet again, there would be further delays until the process was "finished." Never mind what those details were.

I quickly discovered, you will too, that while the astrology is absolute, human free will with regards to sex is strong. The loss of virginity will play out in unpredictable ways. Look carefully and you will see concrete astrological symbolism, but not necessarily in the ways you had expected. This will give you an appreciation of the difference between astrological energies, on the one hand, and the power of free will, on the other. The chart gives you the ingredients. Your free will figures out how to best use them.

Which brings up rape. I would just as soon glide over that, but rape is distressingly common. It horrifies me to realize that many of you have been a victim of rape. As you will learn when you examine charts, men are not always the aggressors nor are women always the victims. Presuming you are learning to use the tools I am giving you, not merely parroting what you read, it will be clear that some people are born with a propensity to rape or to be raped. In other words, there are people whose sexuality is not under their control.

While free will is absolute in absolute terms, individual lives are relative, being bounded by birth at one end and death at the other. Past life studies show, again and again, that unresolved traumas in one life weigh down the next life, until those traumas are resolved, or until time has obliterated them. By "weighing down" I explicitly mean that in those lives, and concerning these explicit areas, free will is limited or does not exist. This is especially true of the first half of your life, when, as you would expect, one can suffer greatly from one's previous life. Rather like a hangover. The best resolution, I suspect, is the second Saturn return. Which I myself have just come through.

If this horrifies you, it horrifies me, too, to know that death does not wash everything away, that children arrive still in the makeup and costume of their past lives, still bearing their scars. If you're curious, this has long been known. In Catholic cosmology, a mangled version of this was called "original sin." In response, the Savior created not one, but two sacraments expressly to deal with it: Baptism, which washes it away generally, and Confirmation, typically administered to seven-year-olds, where they expressly repudiate their own past wrongdoing. I got both of these as a child and, having figured them out, can state categorically *they do not work.*

The power to administer sacraments is given to priests through the sacrament of Ordination, which is said to trace back in an unbroken line to the risen Christ himself. If sacraments do not work, then either the chain was broken (which I think unlikely), or this relic of the Piscean Age has been negated by Aquarian Age, which arrived in 1913. Which declining Church membership over the past century would seem to confirm. When the magic is gone, people leave.

Thus we each bear the accumulated scars of our individual past lives. In terms of raping or being raped, you are looking for malefics in the 5th (bad experiences with sex) or in the 8th (bad life in general), especially when these are in hard aspect to the Sun, Moon and/or ruler of the ascendant, or rulers of the 5th, 7th or 8th.

Consider that actual attacks are the exception, not the rule. Those with distressed 8th houses will likely act out in terms of money and finance. Those with distressed 5th houses will focus on children. Theirs, if they have them, the children of others if not.

Is this the same as sexual preference? In my view there is overlap, but sexual distress and sexual preference are clearly different. As for the claim that one is "born gay" and is therefore helpless, this is absurd. We are all born with preexisting conditions (i.e., conditionally limited free will), as the Church has always said ("original sin") and as the natal chart will prove. We are to make the best use of our lives as we can. In other words, overcome what we dislike, in order to find and enjoy what we do like, in order to more fully enjoy whatever comes beyond life. This is life, in its most fundamental form. The need for a solution to rape and raping is of far greater importance than a preference for partners. See how astrology strips away masks and illusions? Don't take my word for it. Get Hans TenDam's book, *Exploring Reincarnation*. It opened my eyes.

For an example of loss of virginity I will use the natal chart of former president Bill Clinton, a well-known romantic. He was born on August 19, 1946 at 8:51 am CST, in Hope, Arkansas.

I have always had the impression that Bill didn't chase women so much as women chased him — and that he didn't protest very much when he was caught. Clinton's 5th house, the house of casual sex, is empty. Empty houses are areas in which we do not have a great deal of interest.

Losing your virginity

On the other hand, in the 8th house we find Clinton's Moon, which is sensitive to the emotional states of others. It is ruled by Venus, which we find in Libra in the first house. Libra is the sign of balance, the first house is what we project to the world. The result is a man who is visibly (1st house) sensitive to the women around him. Which doesn't seem to make a lover out of him, but, as I mentioned, if you pursue him he won't protest much.

I regret I don't know when Bill Clinton might have lost his virginity (and, yes, he was born with it) as I'm not that good with progressions and solar arcs and whatnot, but the process, how he lost it, is more clear.

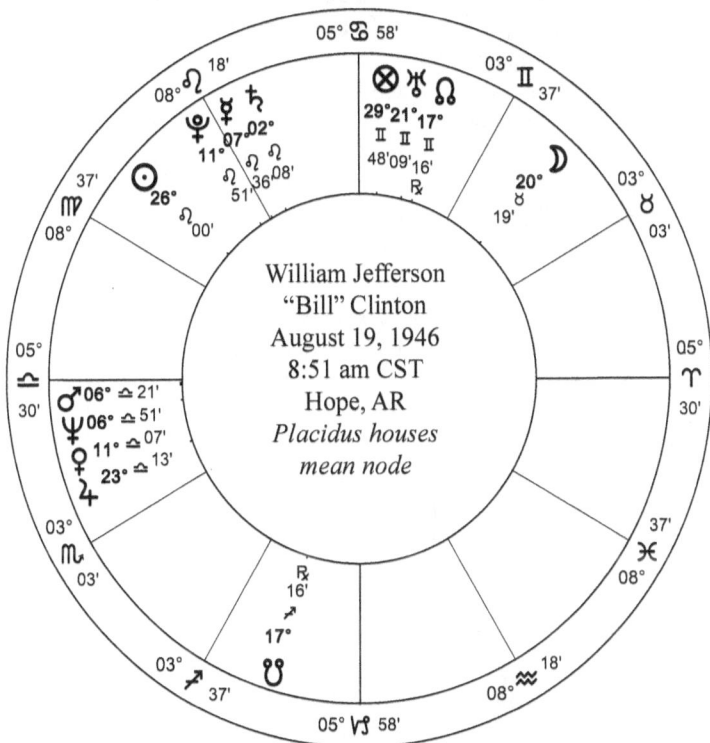

Capricorn on the 4th house, he had to prepare for it, presumably by thinking and wishing and desiring. The 4th, 5th and 7th houses being empty, he had no idea who his partner would be, but here we come to an interesting tweak in Clinton's chart.

Opposite the empty 7th is its ruler, Mars, debilitated in Libra. Opposite the empty 5th is its ruler, the debilitated Saturn in Leo. Next to it is a debilitated Mercury, in Leo. All of these planets would rather be in the houses opposite. Which, if I am not mistaken, gave Bill the fantasy of having sex, of having dynamic (Mars) romantic partners.

In other words, based on what's commonly accepted about Bill Clinton and what is shown in his chart, I am tweaking not only the generally accepted rule about empty houses (that they are not important), I am also tweaking one of my personal rules about debilitated planets, that they want to be in the house opposite, to arrive at this conclusion:

Debilitated planets project fantasies into the opposite houses. The ability to act on these fantasies depends on the chart as a whole. This will especially be the case if those houses are otherwise empty. It is an absolute rule that **if the delineation does not fit the person, the delineation must be changed**, but the result must always adhere to astrological basics. No wild flights of fancy.

This is clear with Bill's empty 7th house.

In Aries, a cardinal sign on an angular house, Bill should have had a series of active, powerful wives who came into his life, turned it upside down, and who then left abruptly, leaving him in the lurch. The house being empty, Bill does not have the final say.

Instead he's only had one wife: Hilary. In Bill's chart, she is ruled by a debilitated Mars in Libra in the first, smack on Bill's ascendant. She identifies with Bill, and Bill identifies with her. They really are inseparable. One would presume that as it's Mars, that they must battle, and as it's the angles, those fights must be in public, but as the sign is Libra (the sign of open warfare, by the way) and as Venus, its ruler, is in Libra and the first, the fights seem to never happen, or to just melt away if they do. That's a powerful Venus.

It's even more clear with the 5th. Saturn and Mercury both want to be in the 5th. Bill fantasizes about sex, about bedroom games, about children, but as a young man did not really know much, the 5th being empty.

So what was Bill's first time like? I think it went something like this:

Capricorn on the 4th says he had planned for it. Saturn ruling from Leo and the 11th, the woman was both unique as well as from among his friends. She was older. Co-ruler of Capricorn is Mars. Mars in the sign of Libra, its debility, makes her a tomboy. At any rate, Bill thought of her as a tomboy, since Mars is square to the 4th cusp. He was not concerned that she was older (might not have known it) since Saturn is inconjunct to the 4th cusp.

As for what actually happened, which is the 5th house, you will note the absence of water signs. This was not a profoundly emotional moment. With Aquarius on the 5th and its principal planets opposite, this was experimental. Try something out, see if it works. The old doctor and nurse game comes to mind, as does "you show me mine and I'll show you yours!" (Yes, I mangled that.)

Overall, it was a easy experience for Bill, since behind it all was Venus in Libra, his chart ruler. He looks so adorable, the girls have always been all over him. Which is to say that Bill's Libra ascendant/Venus in Libra in the first made his private fantasies about sex and girlfriends actually come true. We should all be so lucky.

With this as a start, Bill has continued ever since. It isn't so much that Bill Clinton chases women, but that they chase him, and that he enjoys being caught and tries to get himself caught as often as he can. So far as Hilary is concerned, she lies smack on Bill's ascendant. She has him by the balls, she has physical possession. Where is Bill's heart? That would be his Moon. That's in Taurus in the 8th. The 8th is the house of married sex, not fun'n'games. The 8th, and the Moon, are both ruled by Venus in Libra, and in Libra we find Mars. Which is Hilary.

Yes, there's Neptune in between Mars and Venus. Hilary does have to work to keep track of Bill, but, overall, she doesn't have to work all that hard. She learned long ago that she was the only one he loved, the only one he could ever love, and it might have been the wedding ceremony itself that sealed it.

So far as girlfriends are concerned, all Bill could give them, all he could give Monica, was sex. By contrast, he gave Hilary power, which is 8th house. And as sexy as he is, Bill Clinton is a very powerful man.

As was guessed at the time, Hilary's anger was not that Bill had "cheated," but that he had risked everything the two of them had spent their lives working for: The presidency.

All virile, powerful men separate sex from power. They physically empower their wives, while amusing themselves with maids and servants. Jack Kennedy was the same, and for exactly the same reasons. As single men they seek out a woman who understands this particular quirk, and then marry her so she stays around. The wife gets power and permanency. The girlfriends each get a few good nights, memories for their old age.

You want to know about the impeachment? That was a long time ago. Before I went on and on about rape I had a section on that, but this week's analysis has gone on long enough. The impeachment of Bill Clinton was a showdown between his Leo Sun in a succeedent house, and his Libran chart ruler Venus, in the first. It was never a contest. Venus won easily. Angularity counts for a lot.

I miss Bill. We all do. I wish he were still president.

So, how did you lose your virginity? Was it true to your chart? — *December 27, 2011*

🕐 *The Chart of Charles Baudelaire*
How to Read a Stellium

One of my clients this past week had a major stellium in Leo in the 4th. Stelliums are three or more planets in the same sign, which generally turn up in the same house as well. My client had Sun, Jupiter, Pluto Mercury and the north Node all in Leo in the 4th. Stelliums are read by a distinct set of rules. Rather than embarrass my client, I will demonstrate these rules with the chart of the French poet, Charles Baudelaire.

Charles Baudelaire was born on April 9, 1821, at about 3:00 pm in Paris. He has Virgo rising, with Mercury and Pluto late in Pisces, and Mars, Venus, Jupiter, Saturn and Sun in Aries. The entire stellium is exactly 22 degrees and spans his 7th and 8th houses.

Astrologers have traditionally thrown up their hands with stelliums. They note how close each planet is to the next and which ones are actually conjunct and what this grand conjunction, or series of interlocked conjunctions might possibly mean. In the case of Baudelaire, astrologers note the stellium was in the 8th house and that as the 8th house is about sex, that this explains Baudelaire's preoccupation with sex. Which accounts for the morbid air of his masterpiece, *Les Fleurs du Mal,* the Flowers of Evil (the traditional translation), or sick flowers (my translation).

We read charts by means of dispositors. This planet in this sign and house is ruled by that planet in that other sign and house. Recently I was emailed, or maybe it was a Facebook post, by someone excitedly saying *this planet ruled this* and *that planet ruled that*. Which is a start, but planets and signs, by themselves, have no anchor. They float about like balloons in the wind. To make dispositors work, you need to ground them. You need to tie them down. Tether them, like balloons. Or give them lead shoes: In other words, **houses**.

The first time you try to synthesize a planet in a house and a sign, it's hard. You can put the planet in a sign, OR you can put it in a house. It's hard to do both.

To master planet-sign-house (well, more or less) you need a firm grasp of the underlying planet-house-sign energies. You need to remember that Mercury is not only communication and speed, but also young people of undetermined gender. That Jupiter is not only abundance, but *that which magnifies*. Horary astrology is a good grounding for this, as horary is a practical use of astrology. Once you have this you can transfer one planet-sign-house combo to another planet-sign-house combo. In other words, you can follow a chain of dispositors and you will, at last, be able to read a chart. Learn this and a simple set of rules and you can read anything whatever.

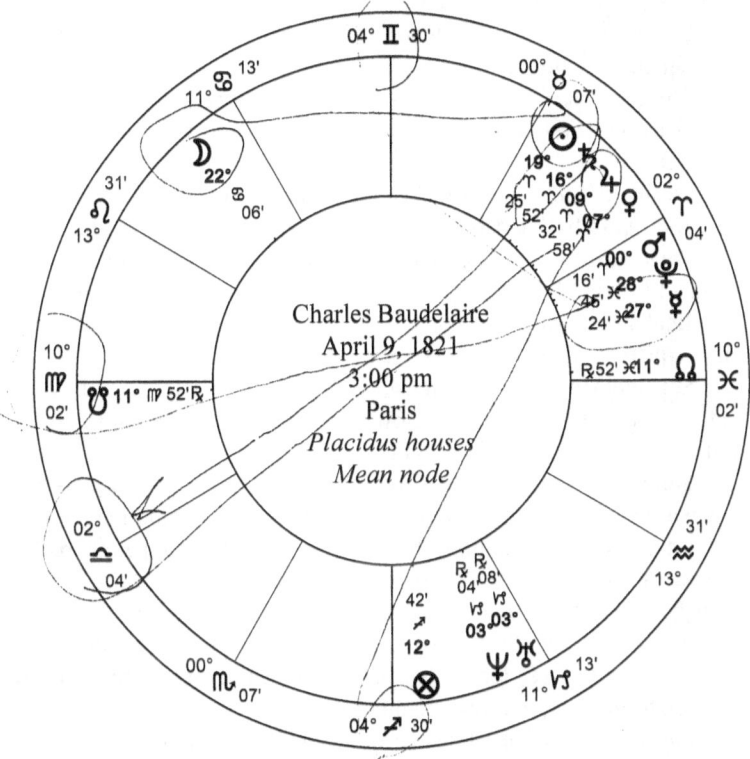

Back to stelliums. Stelliums suck the chart into themselves. Like a black hole. So the first thing you do with a stellium is to note what each planet rules. For M. Baudelaire, here is his stellium, in order of appearance:

Mercury in Pisces rules the MC and the Ascendant. Which makes it a key player. It also rules the north node. Note that it is debilitated.

Pluto does not rule anything, but we will give it an interest in Scorpio, which we find on the 3rd house cusp.

Mars rules Aries, which is to say, Mars rules the Aries part of the entire stellium. Mars also rules the 8th house and 3rd house cusps.

Venus rules the 9th house and the 2nd. It is debilitated.

Jupiter rules the 4th house and the descendant. He also rules Mercury, Pluto and the Pisces part of the stellium. Because in his turn Jupiter is ruled by Mars, all seven planets and both signs are part of one unified grouping.

Saturn rules the 5th and 6th houses, as well as Uranus and Neptune. Saturn is debilitated.

The Sun rules the 12th, and, with Mars, co-rules the 8th house cusp and the Aries part of the stellium.

Elsewhere we note the Moon is in Cancer, which it rules, and is placed in the 11th house.

There is a very exact Uranus-Neptune conjunction at 3 Capricorn. Both retrograde, they are trying not to be in the 5th house and are stranded as a result.

To sum up, in Baudelaire's chart we have three planets in dignity: Sun, Moon and Mars, and three in detriment: Mercury, Venus and Saturn.

We look at Sun and Moon, because the Sun and Moon, for better or worse, will always be key players. We find them both in dignity, and both in cardinal signs, which tells us this is a powerful individual. They are tightly square, which is an indication of stress.

Sun and Moon are consciousness. Sun is what's going on in your head, the Moon are your feelings. So we start the interpretation by building the basic Sun-Moon relationship:

Sun in Aries is headstrong and impulsive. Moon in Cancer is hypersensitive. This already does not look like a good combination. Sun in the 8th has to do with the nasty bits of life and death. Do people really like you? Do they trust you? Do you trust them? With your life? Aries makes it all or nothing, in particular with accidents (Aries) or inflammatory illnesses (burning fevers: Aries) as these can be life threatening.

Moon in 11 lives for his friends, but, when stressed, will demand too much from them, thus driving them away.

Put Sun and Moon, Aries and Cancer, 8th house and 11th house together and we have someone who will brutalize his friends. How much will he do this? Precisely in proportion as the square aspect is exact: The orb of aspect. So far as orb vs: whole sign, I think it works like this: Within the orb, one is conscious, more or less, of the aspect and what it can do. Outside of the orb, but within the signs themselves, one expresses the aspect but is not aware he is doing so. Charles Baudelaire's Sun and Moon are less than three degrees from exactly square. His demands on the people around him, that they prove themselves worthy, are conscious. As is his frustration when they do not or cannot.

How to read a stellium

We turn now to the biography. Baudelaire had a passionate love for his mother. Why? The mother is the 10th house cusp, which is Gemini. The ruler is Mercury, which is in Pisces and the 7th house.

Gemini as the mother, she is two-faced, like Gemini itself. On the one hand she has her proper life (husband, second husband), on the other she is weighed down by her son, represented by the ruler of Gemini, Mercury, which, among other things, represents children. I mean "weighed down" literally. Mercury is debilitated in Pisces. Consider it as the mother, she is an emotional, slobbering, immature mess. Considering Mercury's house placement, which is as Charles's partner, we see that Charles expects an emotional, slobbering mess. Considering Mercury represents children, and that it is Charles' chart ruler, we can see that Baudelaire himself is an emotional, slobbering mess.

Note this: Planets are like actors. They wear many hats (sign placements). They take on many roles (house domiciles). Are you a fan of Cary Grant? Watch his old movies on *Turner Classic Movies*? Like to see what he can do with a given character? Or are you a fan of Paul Newman? A great actor has great expressive range. Anyone who tells you that Saturn — or the Sun — is your father, or that the Moon is your Mother, etc., is in astrological kindergarten. Planets take on different roles depending on what you are asking of them, how you are studying them. Baudelaire's Mercury simultaneously represents both himself, and his mother. In your chart Mercury will represent something entirely different.

Mercury's ruler is Jupiter in Aries in the 8th. The 8th is the partner's money. Baudelaire expected his mother to give him money (lots of it: Jupiter), but, mommy, who is, remember, Gemini on the 10th, has no money for her son because her son's 10th house is empty. What she has she won't give to her son, because the ruler, Mercury, is debilitated in Pisces. Debilitated means that, given a choice, it won't do the right thing, and, Pisces looking to Gemini, will feel justified in lying about it. Gemini lies deliberately, Pisces is too unconscious to know what the truth is, and as a result prays to Jesus, the Piscean God, for guidance, all the while being duped and swindled. After her son was dead Mama Beaudelaire paid his debts, which is penny wise and pound foolish. As Baudelaire himself had complained.

And that is an overview of the fundamentals : Sun, Moon, ruler of the ascendant. Baudelaire in a nutshell. Let's go further.

Why is Baudelaire a poet? The classic poet signature is a Venus-Mercury conjunction, in either a sign of Mercury, or a sign of Venus. Beauty (Venus) in words (Mercury), especially when they fall in the 3rd. Note the conjunction is the only significant aspect the two planets can make.

In Baudelaire's case, Mercury in Pisces is ruled both by Venus and Jupiter, conjunct in Aries. When Aries owns Pisces, as it does so far as Mercury is concerned, it will drive it. Here, Jupiter (volume) is superior to Venus (beauty), since Venus is debilitated. Venus wants to be beautiful, wants to be in Libra and the 2nd, but isn't, and so becomes ugly instead.

The "ugly" in art, which we see so much of today, seems to have been an accidental result of Goethe, who invented the Romantic movement. Goethe achieved this through his full moon (Virgo-Pisces), which gave him an intense, broad range. Lesser talents mistook broadness for simple shock value, which is, sad to say, ugliness. Because ugliness is hidden and because whenever it comes to light it is shocking, we tend to confuse ugliness with the truth. In reality, what is true is liberating and uplifting, whereas ugliness coarsens us and makes us cruder in mind, emotion and spirit. Ugliness is at first shocking, then challenging, and finally commonplace. Good art requires great skill. Bad art merely requires boldness. Left to itself, ugliness will win in the end and civilization will be lost, until we can no longer imagine anything better than what we have. Please do not ask my opinion of popular culture.

With Baudelaire, with certain late works by Beethoven, we see the start of ugly.

But we can go further with Baudelaire. What kind of ugliness will he give us? Mercury's ruler in the 8th house, we get intensity. Baudelaire will keep going until we are shocked, one way or another. Remember his Sun and Moon are themselves stressed, so he presumes the world is like him, that it is stressed, too.

But intense does not mean ugly, nor does it mean sex. We get both from Pluto's conjunction with Mercury, as would be expected. This is picked up by Venus, debilitated in Aries. Venus in Aries wants to be in Libra, it wants to have a partner, it wants his attention. If Venus were in Libra, it would bill and coo and charm us. In Aries it takes a militant approach, attempting to reach that same end. But instead we are shocked.

It is said that in his teens Baudelaire contracted both gonorrhea and syphilis. Which are sex diseases. As these are transformative (they will kill you), they are 8th house, not 5th house, afflictions. Run a transit and you will find Pluto going over Baudelaire's Saturn around 1838. Pluto carries disease from the partner (7th house) into Baudelaire's 8th. Pluto infects Saturn, who, debilitated, cannot resist, cannot save himself. Going back a few years previous, Pluto ran over Baudelaire's Venus around 1829, when he was about eight years old. Which is presumably when he got his first sexually transmitted disease. This is very young for sex, but it was the year after Baudelaire's mother remarried, which he took as a sign of rejection.

The 8th house has Aries on the cusp, which makes it impatient for sex. Mars is in Aries but just outside. It will impulsively want to get into the house, to get started with real sex, and, in particular, seize Venus, the very first planet actually in the 8th. Debilitated in Aries, she is his hostage.

So long as Baudelaire could hope that his mother would be his partner, Mercury, both the chart ruler (Baudelaire himself) as well as his mother's ruler, would keep him innocent, keep him hoping. It appears that when she "abandoned" him to get on with her own life, he acted out. Venus debilitated in 8, ruling the 2nd, he paid for the encounter. Baudelaire was known for his associations with prostitutes. Which for Baudelaire was 8 and 2. In other charts it will be 5 and 2.

What about Uranus and Neptune, conjunct in Capricorn, just outside the 5th house cusp and wanting to run away from it?

Saturn, their ruler, in the 8th house, and Mars, the co-ruler, also in the 8th, will whip or bully them into compliance. Get them into the 5th where they can have fun with them. The result is kinkiness.

Which reminds me that the Marquis de Sade had his Sun and Moon in tight square (Gemini-Virgo), with Saturn debilitated in Cancer opposite Uranus in Capricorn. (See *Duels at Dawn*, pg. 115.)

If I knew more about Charles Baudelaire I could go further. Somebody who wanted to write a killer expose on, say George W. Bush and wanted to hire me to work out his chart, in close study of the events of his life, would produce stunning results. It is when you really know a man that you understand the artifacts which he unwittingly scatters behind him. It is when you know a man that you will know where to look for his secrets, and what they mean when you find them. It is for this reason that astrologers must forever remain silent about their clients, as there is literally nothing a skilled astrologer cannot know. — *January 3, 2012*

☉ Daffodils and Haircuts

I went out on New Year's day and found the daffodils were already up. Today they are 2¾ inches. They will bloom before the month is out. You know you're having a mild winter when the daffys peek out at the very beginning of it. Is this "global warming"? Well, maybe. The winter of 1960 was so warm my second grade teacher's bulbs bloomed early. What do the *Old Farmers* say? Let's look at Chicago. That's pretty much the center of things.

For 2012 in Chicago, winter will be *"slightly milder than normal."* One of the coldest periods will be in late January. So far, it's been a lot milder than "slight". For the week of January 8-12, Chicago's forecast is "snow showers, cold." Today (Sunday), Chicago is forecast to be four degrees above average. 40°, instead of 36°. That's not "cold." So if the Old Farmers say "cold" and it's mild, then it really is mild. On the other hand, the little bar graph on the bottom of pg. 207 shows temperatures to the max for January (+4°) and then a fall, to the max (-4°) for February.

The Farmers say summer in Chicago will be cooler and drier than normal.

I get haircuts about twice a year, and while I may look shaggy most of the time, I pick the day carefully. I've been using the Matrix Electric Almanac for years. It was a simple DOS program that was freeware. Matrix gave it away and urged others to do so as well. It's now on-line and still free. Are you in the beauty biz? Run sales, run specials on the best days, but also *close* the day before, and the day after, as those are real stinkers. Do this for six months and you'll have all the heads in town. Matrix got their stuff from Robson's *Electional Astrology,* but the Matrix format is easier to use, and you don't need birth data. *— January 10, 2012*

☉ *The chart of William Shatner:* **The Sun and Moon**

A week ago I was going to give you William Shatner's chart, as a prelude to his time twin, Leonard Nimoy. But funny thing, when I looked at Captain James T. Kirk's chart (born March 22,1931 at 4:00 am in Montreal Est, Canada), I found I could not read it.

Aquarius rising, ruling planet Saturn in Capricorn in the 12th house, Sun-Mercury-Uranus-North Node all in Aries in the 2nd, Moon in Taurus in the third, this was not in any way an actor's chart. This is a guy who wants money (2nd house stellium). Moon in 3, a guy who wants to be a big cheese, provided he doesn't have to work too hard at it.

Shatner in fact holds a Bachelor of Commerce degree from McGill, exactly as one would expect. So why is this man an actor and how did he get that way?

My guess is that some mysterious older person — chart ruler Saturn in the 12th — pushed him in that direction. Since this is a 12th house matter I suspect we will never know who, but my guess would be a maternal aunt or uncle, since the 12th is the third from the 10th. The 10th is the mother, the third from it is her brothers and sisters, which is Shatner's 12th. That this person was an outcast is hinted from Saturn's opposition by Jupiter, Pluto and Mars, all in Cancer in Shatner's 6th. These would be his father's brothers and sisters. You can have lots of fun with charts by "turning" them like this. If I believe Robson, you will often actually be correct. If not, your hunches will eventually be thrown back at you and then you will know. We learn best by dare and risk.

Shatner, it turns out, just wants to work, just wants something to do. Give him a job and he's happy. The worst period in Shatner's life was after Star Trek was cancelled and he was very nearly homeless. Since then he has played a variety of blue-collar roles. He will be 81 years old this spring and is currently grinding out commercials. Which, as anyone in the industry will tell you, is the very bottom of Hollywood. To add insult to injury, he persists with shabby jobs decades after his contemporaries have retired. Bill doesn't need the money, but good luck convincing his packed 2nd house of that. He is busy because his Moon in 3 makes him busy. Money is what he gets in exchange. Houses two and three, Sun and Moon. Such is Bill Shatner. The charts of one-dimensional character actors may be just like Bill's: someone who makes a living by acting. There is a drive for success and a need for money, but no specific talent for the work itself.

I have been asked for my "technique," my "method" for reading charts. Since I am stream of consciousness, I never sort out mechanical details until I have to. So I have thought about it.

We are known by the ascendant and by the house and sign placement of its ruler. Myself, for example. I have Gemini rising with a strongly placed Mercury in the 9th house, in trine to the ascendant. People think I "know things," and generally I do. In Shatner's case, we *don't* know him because his ruler is in his 12th. So we pick up on his Sun instead. Betcha you always knew that Bill was an Aries.

On the other hand, we know ourselves, not by our ascendant nor its ruler, and not by the sign the Sun is in, but by the *house* the Sun is in. The *house* of the Sun is what we are *conscious of.* The *sign* is how we *express that.* Shatner, for example, is driven by a *need for money*—his second house. He gets money by being an Aries. For Shatner, Aries is a *tool.* Money is a *need.* Signs are tools. Houses are needs.

Shatner values himself by how much money he has. When he lived in the back of his camper in the early 1970's, he did not feel he was starting a new Aries adventure, he did not believe himself to be Boldly Going into a new rustic life. He was depressed because he was bust. His tool, Aries, was not providing for his needs, which, second house, was a need for money.

Shatner's chart is a useful for another reason, in that the second house, the house of values, is a useful introduction to house rulers and dispositors. Modern life comes down to sex and money, and while you can never be quite sure exactly what someone's sexual cravings may be, a glance at their lifestyle will tell you everything you need to know about their money. How much, where it comes from, what they do with it, etc. The second/eighth cusps will tell you.

Some tool/need combinations work better than others. For example, my Sun is in the 9th. I *need* to know things, I *want* to know things, I am happiest puzzling things out. Big philosophical, religious, foreign things, since this is the 9th, not the third. But I am *unhappy* with Sun in Aquarius as my tool. Aquarius marginalizes me, pushes me into obscure and unimportant topics, makes me subject to popular whims and fancies. The raw intensity of Scorpio, the hard work of Capricorn, those are more my style.

Sun and Moon: William Shatner

Surprised? Are you happy with your Sun sign?

If the Sun is consciousness, the house and sign of the Moon is where you are most sensitive. The Moon is a lot trickier than the Sun. The Moon has a thousand shades of happy and unhappy. It will tolerate some things almost forever, but refuse to budge with others.

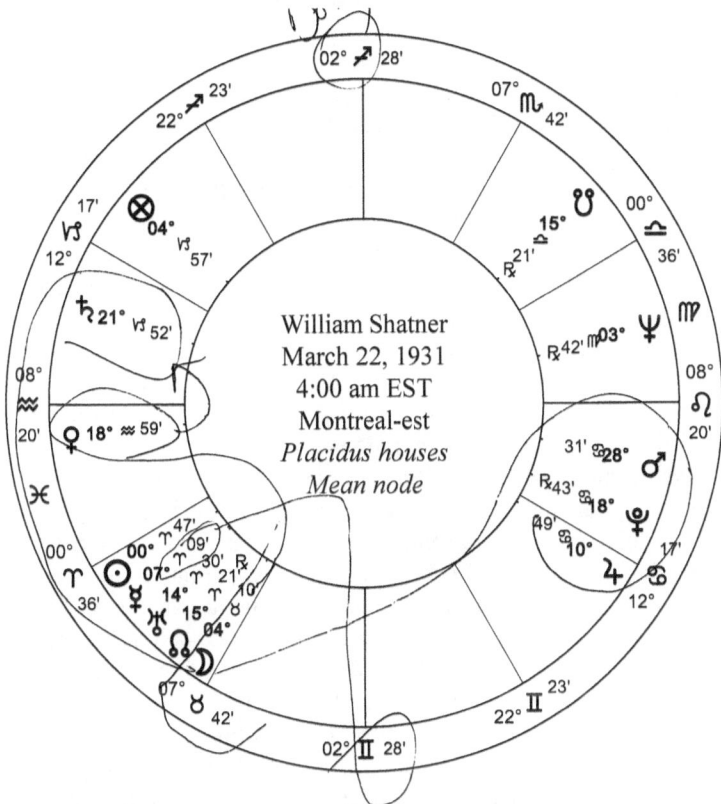

The Moon tends to accept where it is, its house, that is. The 1st, 7th and 10th houses tend to expose it, as does the 3rd and 11th. The 4th and 12th conceal it. The second makes it needy, the 6th, sickly, the 8th is too intense in general, etc. Which leaves the 5th and the 9th, which strike me as neutral.

Next, consider the Moon's element and aspects. Earth signs, for example, are soothing. You can pile a lot of stressful aspects on an earth Moon and it won't object. Water is a variable. Cancer is touchy, Scorpio is intense and nasty, Pisces is unfocused. Air is restless and, I would think, lacks depth. Fire is dynamic, but when stressed can explode.

In the case of William Shatner, Moon in Taurus will make the day to day life of the 3rd house go quite slowly. A lot of it will pass him by, since

Taurus can be quite slow and the Moon will not be in any hurry. In fact Shatner lives on a ranch ESE of Fresno, where he raises horses. In puzzling out why a man with Gemini on the 4th house would live on a ranch, we see how clever the Moon can be in satisfying her needs.

Shatner's Moon in 3, but in Taurus, daily life (3) should be connected to the land (Taurus). But home, which is Gemini, should have some aspect of motion, of mobility.

Such as can be given by, let's see.... *horses* — ?

Horses, "great animals" to horary, are ruled by the 12th. Which is a "natural affinity." Rather than an "accidental" one, which would be specific to Shatner's chart (the ruler of 12th house (horses) in Shatner's 4th (home)). Instead, we find Shatner's chart ruler in the 12th, which is Saturn. Which is to say that when William Shatner "goes home" to his chart ruler, the natural seclusion of the 12th is conducive to breeding fine horses, as Capricorn will work until perfection is achieved. It also gives Shatner the privacy he craves, as I can find no photos of his ranch. You will note that although the Moon does not like Capricorn, it finds the trine to the 12th house, and Saturn, to be comfortable. You will also note the trines from his Moon to Jupiter, Pluto and Mars, all in Cancer, which the Moon rules. These trines give Shatner's Moon great depth.

Also note that the ruler of Shatner's 4th is Mercury, in the 2nd house. Shatner will want to make money at home. He does so by raising horses.

As this is convoluted (horses in 12 in a general way, the need for "something that moves" to make the 4th happy, needing a daily routine that is slow and deliberate, making money from home, wanting seclusion, etc.), it is not surprising that it was not until the later half of his life that Shatner, born and raised in a city, could give himself a rural life.

What is simple and direct — and satisfying — can be done at an early age. Which the chart will show. Convolution takes time and will produce surprises. Dispositors give us a great deal of information, but when we study actual people we realize just how complex they can be.

As a rule, when the Moon is happy, we are happy. Shatner was lucky. While his Sun and Moon are not in any way related (not by dispositor, not by aspect), his Moon has comforting aspects and Shatner was eventually able to give it a ranch to play with.

Not so lucky was Goethe. Like Shatner, Goethe has a 3rd house Moon, but unlike Shatner, it is in unfocused Pisces and opposed by the Sun. The third house is one of the houses in which the Moon is "exposed." When Goethe walked about the streets, when he ran errands, when he was a student in school, his Piscean Moon picked up everything that was out there. A lot of which wasn't very nice, but none of which he could tune out. Pisces is defenseless. What survived and made it home was then held to critical account by his Sun in Virgo.

So you will not be surprised that in most of his portraits, Goethe appears frightened. The typical result of a 3rd house Moon is seclusion. I myself have a 3rd house Moon and am rarely seen in public.

One of the most critical factors for the Moon is its aspect with the Sun. A trine or sextile from Sun to Moon, from head to heart, can make life flow, give you an ease and grace and self-confidence. Sun-Moon trines tend to form between compatible houses. Vedic astrology calls houses in trine Trikona ("trik") houses. Houses in trine tend to relate to one another. There are four sets:

1, 5, 9 (individual, happiness, education);
2, 6, 10 (money, health/work, career);
3, 7, 11 (brothers, sisters, spouse and friends);
4, 8, 12 (home, occult).

When your Sun and Moon take up two of these, you have a natural focus for your life. When Sun and Moon are in sextile, you have the Moon in one set, the Sun in another.

Sun and Moon in square — first or third quarter — is an aspect of stress. They try hard. They have an "itch" that must be scratched. Again, the houses they occupy will dominate the chart. In most cases Sun and Moon will both be in angular houses, or both succeedent, or both cadent.

Sun and Moon square in angular houses (1, 4, 7, 10) are outgoing. Sun and Moon in cardinal signs and angular houses will be strong and self-confident. In fixed signs they will be visibly stubborn. In mutable signs they will be easily — and publicly — led.

Sun and Moon square in succeedent houses (2, 5, 8, 11) are conservative. They want to hold on to what they've got. If in cardinal signs, they are active. If in fixed signs, lazy (why change?), in mutable signs, grasping, as they will tend to hoard, if I am not mistaken.

Sun and Moon square in cadent houses (3, 6, 9, 12) need direction, but don't always get it. In cardinal signs they will follow orders. In fixed signs they will seek stability, in mutable signs they are chaotic.

Sun-Moon conjunct will intensify the house they are in, by means of the sign. This is an aspect of single-mindedness. These people are serf-absorbed. If Sun and Moon are both in the second, for example, life is all about money. In the third, they are busy, in the 6th, work and health and diet, in the 7th, partnerships, etc. The sign tells you how this is done. If Capricorn is on the second, money must be earned. If Leo is on the 7th, the partner will be proud and perhaps childish, etc.

Whenever you see Sun and Moon conjunct, look immediately for the node (north or south) because if it is within 18°, they were born on the day of an eclipse, which adds a special degree of intensity. The closer they both are to the node, the more intense the eclipse.

Sun opposite Moon tends to polarize the houses occupied. This is the most dynamic and powerful of all aspects, as Sun and Moon will literally hunt each other down and pound each other to bits. What they produce will often deliberately shock. Again, look for the node to determine if there was a lunar eclipse.

Sun and Moon inconjunct, an aspect of invisibility, do not "see" each other. The head does not know what the heart is doing. These two will connect through their dispositors, and if the Moon is in Leo or the Sun in Cancer, they will connect with each other in strength.

It sometimes happens, as with Mr. Shatner, that the Sun and Moon are not in any aspect, but the urge is so strong that a convoluted chain of dispositors may eventually emerge.

You can learn a great deal about yourself, a great deal about others, by studying the life (not the charts, per se) of those who have the Sun and Moon in the same houses as yourself. Which in many cases will also be the same aspect.

You will tend to dislike or distrust those who have Sun and Moon in the opposite houses from yourself (Sun in the opposite house from your Sun, Moon in the opposite house from your Moon), while you will, most likely, be attracted to those who have the Sun and Moon reversed (their Sun in your Moon's house, their Moon in your Sun, etc.). This will be especially true if the signs are compatible. Full moons get a double-whammy, since those who have the Sun in the opposite house will have it in the same house with the Moon. To my knowledge, my full moon has yet to meet such a person, one with a full moon in inverse houses to mine.

So far as reading a chart, get the Sun and Moon right and you've got the heart of it. The other planets, the other signs, are extraneous. Note carefully how they relate, not so much to each other, but to the Sun and Moon. Above all, to the Moon. — *January 10, 2012*

☉ Whole Sign Houses

I was asked today about whole sign houses. I in fact use whole sign houses, but not the simple sort. The simple sort of whole signs are 30° chunks counted off from the rising degree. These fail for many reasons, chief among them the failure to find a midheaven, which we know to be a significant point. Why does Vedic astrology use these simple houses? Because India is very nearly equatorial. In such climes houses are very nearly always equal no matter how you divide them. As we go further north (or south) from the equator, the ascendant-MC angle can reach 120 degrees or more. Europe, for example.

Which gives us quadrants which must then be divided. There are many ways of dividing. The simplest is to trisect zodiacal longitude (e.g., from Aries to Cancer), which is Porphyry. Which isn't really a house system at all since these do not relate to the earth itself. Proper division, with relation to the horizon and various great spheres, gives intermediate cusps which, by experimentation can be shown to be sensitive to various things.

My modified whole sign system says the sign and the house must agree, but degrees in front of the cusp must be somehow be of lesser value than degrees properly inside the house. Thus, "running to get in."

Note that when a sign does not have a cusp to call its own, it has no way to express itself in the chart. Planets so posited can be "lost" and become sources of frustration and can act out. This happens for two reasons: First, when an entire sign is intercepted inside a large house. Second, as angles have no orb on their backsides, planets in cadent houses, but not in the sign on the cadent cusp, are stranded. A 9th house Sun in the same sign as the MC, is stranded in the 9th. It is not in the 10th. It acts out. — *January 17, 2012*

⊙ Newt Gingrich HATES Mitt Romney

Right now, I don't think anyone cares who the next president will be. It's clear it won't be Ron Paul, who, despite grave faults, would at least be amusing. The next president will, at the very least, pander to the rich—as if he were a mere toady—while using the office to act out his more lurid fantasies. The United States of America has become an elective dictatorship. In such times actual governance is a mere accident.

I used to wonder when a president would simply refuse to leave, when he would declare himself President For Life, but then I remembered that most such leaders end up murdered. The American system is, in fact, ideal. A Pretender struts to power, gorges himself for four or eight years, and then retires to live the rest of his life in obscene luxury.

American presidents have had little success in hand-picking their successors. There is a reason for this, too. Unlike in, say, Mexico (which has, in fact, shot some of its former leaders), former US presidents do not fear their successors will have them arrested and tried for crimes committed while in office. Whenever the public howls for the head of the person responsible, a sacrificial lamb is selected from among the hirelings.

All through 2011 we were treated to endless Republican debates. Again and again we would see a line of pretenders. Always a line. Never a circle. It was clear they had no interest in forming a daisy chain, despite the pleadings of 300 million bored Americans who desperately wished they would. You're thinking of that now, aren't you? Herman Cain and Michele Bachmann, in particular.

Which led to a game of musical chairs. Those who play musical chairs do, in fact, make circles and as the seats got ever more scarce, it was a matter of time, I suppose, before musical chairs turned into circular Russian roulette.

Which happened a couple of weeks ago, when Romney turned massive firepower on Gingrich, virtually tossing him out of Iowa, leaving a dumb-

Newt Gingrich Hates Mitt Romney

founded Santorum standing. With an emphasis on dumb, by the way.

Gingrich has now struck back, with a customized, highly damaging video assault on Romney which he is airing in advance of Saturday's South Carolina primary. I haven't seen it, but I understand Newt correctly exposes Mitt as being self-centered and greedy. Not a "job-creator," but a job-destroyer. *Dem's fight'n words.* They have lifted this dull campaign out of the doldrums and into a nuclear exchange.

Those who have watched Newt over the past 30 years know he can be a nasty fighter. Is his heart in it? Let's compare Newt and Mitt and find out. Follow along:

Newt Gingrich was born June 17, 1943, at 11:45 pm EWT in Harrisburg, PA.

Mitt Romney was born March 12, 1947 at 9:51 am EST in Detroit.

There is good stuff between the two: Newt's Mercury is conjunct Romney's north node and ascendant, trine his Venus and MC. Newt's Jupiter trines Mitt's Sun, Moon and Jupiter. We often find good aspects between enemies, because without them there would never have been anything to bring the two to each other's attention.

A handful of good aspects is often the setup for unhappy ones. Romney's Sun is square Newt's Sun and Moon. Since Newt is a full moon (Sun in Gemini, Moon in Sagittarius), this puts Mitt's Piscean Sun squarely in Newt's sights. Full moons tend to polarize and as a result are often keenly aware of the signs in square to themselves, as signs in square can be anything from a breath of fresh air to a constant irritant.

In this specific case, Newt's full moon, from Gemini to Sagittarius, as well as the signs in square to it, Virgo and Pisces, are all ruled by Mercury and Jupiter. Which makes squares to Newt's Sun and Moon particularly touchy as the ruler of the planet(s) in square will naturally favor one side of Newt's full moon over the other. Either Mercury or Jupiter. Which upsets the balance that full moons, in general, spend so much time trying to achieve.

Is this true of all the polarities? No, it's not. Leo-Aquarius, for example, are ruled by Sun and Saturn, but the signs in square to them, Taurus and Scorpio, are ruled by Venus and Mars.

So, since Mitt's Sun in Pisces falls in Newt's first house, Mitt is a player in Newt's chart. The books say that when a friend's sun falls in your first house, this is an aspect of mutual identity between the two, regardless of gender. Drinking buddies can have this, however, in Newt's chart this is not so simple.

In Newt's chart, Pisces and Virgo are intercepted, in 1 and 7. Mitt's sun falls into Newt's intercepted Pisces. The exact location of interceptions depends on the house system you use. While every house system that produces an actual ascendant and a real MC will produce identical quadrants (houses 1, 2 ,3; houses 4, 5, 6, etc.), different house systems divide those quadrants differently and will produce different results. Placidus will give different interceptions than Koch, or Porphyry, or Regiomontanus or Campanus, etc. Which house system is "right" is a very, very

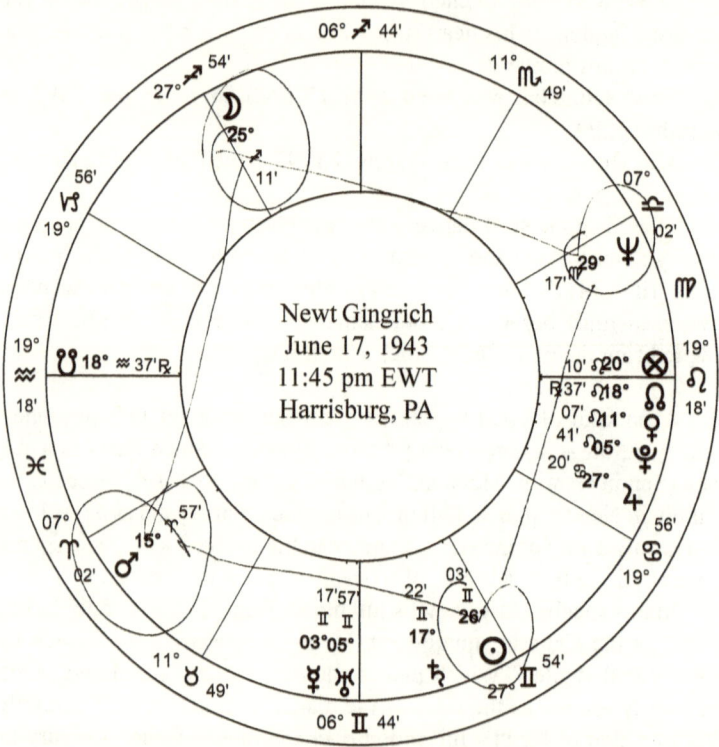

sticky subject. I have elsewhere suggested that space based systems are "static" and suitable for questions (Regiomontanus/horary), while time based systems, such as Placidus, are dynamic and suitable for living organisms, in other words, charts that exist through time, but this is only a crude distinction. (See *Duels at Dawn*, pg. 92.)

Koch, for example, is a member of the time-based systems, but gives very different results from Placidus. My own experiments, many years ago, convinced me that Placidus was better than Koch, so far as reading a chart went. But Dr. Koch never claimed his system was intended for delineation. His claim was that Koch houses, the intermediate house cusps

(2, 3, 5, 6, 8, 9, 11, 12) were sensitive to simple transits. Which they are. One of *Dave's Rules* is to use the house system that is appropriate to what you're doing.

So when Mitt's Sun pops into Newt's intercepted Pisces in his first house, it sets up a confrontation between Saturn, the ruler of Aquarius on the ascendant, and Jupiter, ruler of intercepted Pisces. In Newt's chart, these two planets are square by sign. Squares are forms of irritation, but,

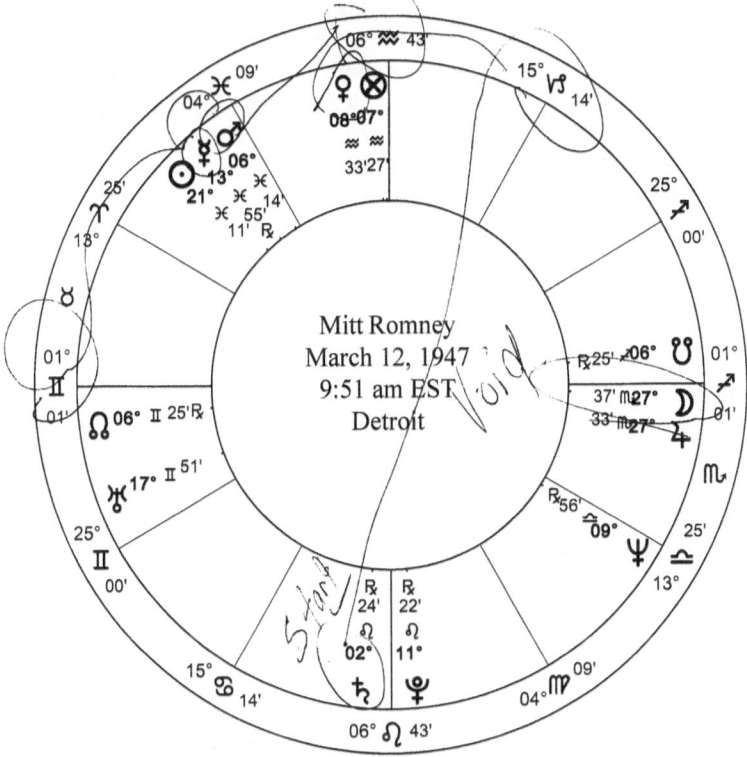

ten degrees out, the irritation is normally minor. Romney's Sun, because it squares Newt's full moon, not only pulls Gingrich off balance, but also brings his Jupiter-Saturn square to the fore.

I will be frank and admit I do not remember all my basic delineations off the top of my head. When I need to cheat and look something up, I go to Sakoian and Acker's *Astrologer's Handbook*. Where I read that Jupiter and Saturn in square are missed opportunities. Shooting oneself in the foot. Which Gingrich has done countless times in the past. (It's why he's no longer Speaker of the House, for example.)

Which maybe explains why Gingrich went nuclear with Mitt. In advance of Iowa, Mitt turned heavy firepower on Gingrich, shot Newt's presidential hopes to pieces, and in the process reminded Gingrich of his own personal failures. Whereupon, caught by the short hairs, Newt dropped his obsession with repealing child labor laws, and lashed out. Viscerally. PS. "Short hairs" refer to the hair on the scruff of the neck. Not those other ones!

By the Lord Jehovah, you do not want Newt Gingrich as your personal enemy. He, more than any other man alive, is personally responsible for the polarization in the world today. In revealing Mitt Romney/Bain Capital as a bunch of money-grubbing job-destroyers, Newt is accidentally exposing the Reagan fraud of trickle-down economics which have dominated US politics for the last 30 years and which Gingrich himself is partly responsible for. When Newt Gingrich is angry, when he has you in his sights, he is punch-drunk. He doesn't care if he fires on one of his own. If you have offended him, if you have injured his ego, there is no mercy. He would make a horrible president for this reason alone.

You will recall his years-long campaign to impeach Bill Clinton. Before he could achieve success he himself was held to account for his own romantic irregularities and removed from office, a blow from which he has never recovered. One wonders if the Clinton people did not take Newt aside early on and attempt to blackmail him (politics is not a nice sport), only for Newt to shrug them off, to his own peril: Jupiter/Saturn square.

The last few days I have heard that "of course" Newt will call off his attack and not harm the Republican party, but that does not sound like Newt. "Penny wise, pound foolish" is his style. Or as Goethe, another full moon, is reputed to have said, Be bold. Boldness is its own reward. Newt has never been afraid of that, and boldness, or brazen indifference, is, indeed, often an outstanding feint.

How did this look to Romney? Rather more severe than you might think. Since Mitt sets off Newt's full moon, we start by putting Newt's full moon into Romney's chart. We are surprised to find it lands exactly on Mitt's 2nd and 8th house cusps. Which makes not only Romney's money an open book, but Mitt's sense of his own values. In Romney's chart these houses are largely empty (Uranus is a minor player to me) which means that Newt can make a very painful fuss if he wants to.

Since we are now talking of Newt's chart "invading" Romney's, note the placement of the rulers of Newt's full moon in Romney's chart. Newt's Mercury falls almost exactly on Romney's ascendant. This can be read as Newt's ideas coming to Romney, which are then fed back to Romney's own Mercury, which happens to be double-debilitated, retrograde in Pisces, sandwiched between Mars and the Sun, both in Pisces. This is an especially

Newt Gingrich Hates Mitt Romney

badly placed Mercury, as it is not only weak, but is disposed by Jupiter in a rudderless Scorpio in the 6th house, which is cadent.

Which is to say that not only has the vastly more intellectual Newt demolished Romney (Romney isn't that bright), but the very personal attack that Newt has mounted has very likely put the fear of God into Mitt. Yes, even Mormons have a God.

In response, Romney's Moon might come to the fore. Romney keeps it under wraps mostly as he's never figured out how to let it out without appearing to be quite nasty. Moon in intense Scorpio with Jupiter sitting on top of it won't be very nice.

Doesn't Romney remind you of Otto, from *A Fish Called Wanda*? And that is such a wicked image I cannot resist:

Wanda: [*after Mitt breaks in on Wanda and Archie in Archie's flat and hangs him out the window*] I was dealing with something delicate, Otto. I'm setting up a guy who's incredibly important to us, who's going to tell me where the loot is and if they're going to come and arrest you. And you come loping in like Rambo without a jockstrap and you dangle him out a fifth-floor window. Now, was that smart? Was it shrewd? Was it good tactics? Or was it stupid?

Mitt: Don't call me stupid.

Wanda: Oh, right! To call you stupid would be an insult to stupid people! I've known sheep that could outwit you. I've worn dresses with higher IQs. But you think you're an intellectual, don't you, ape?

Mitt: Apes don't read philosophy.

Wanda: Yes they do, Otto. They just don't understand it.

Which is to say that Newt Gingrich, ace Washington infighter, has demolished Mitt Romney. Gingrich might have accepted losing, but when Romney, his intellectual inferior, singled him out in advance of the Iowa caucuses, Newt took the gloves off. DO NOT underestimate the sheer power of a full moon birth.

So how does this look to the rest of us? Let's peek at the election day charts themselves.

Primary day in South Carolina will be Saturday, January 21. Polls open at 7:00 am, the capital city is Columbia. A chart set for those coordinates gives 21 Capricorn rising. *When reading a chart it is crucial to know what you are reading and why you are reading it.*

The chart for when the polls open show the voters and what they are voting for. The candidate who best expresses those hopes and wishes will be the winner.

Capricorn rising is a dutiful electorate, not a happy one. Ruler Saturn is in Libra in the 9th house. The 9th house in an American chart is about religion, but also about strangers. Mitt is a Northerner and a Mormon, which is a strange beast down that way. It ought to be easy to run against a New York financial whizz kid and former governor of Massachusetts.

Saturn's sign, Libra, agrees with the sign on the cusp of the 9th, which is also Libra. Libra is ruled by Venus, which we find in Pisces in the second house.

Venus in the second says the voters will go to the polls to vote their pocket books. They want more money! Newt has labeled Romney as a job killer. No one is going to vote for that. No one.

And they know it: Note how Romney's Mars is right on top of election Venus? Like an unwelcome suitor, too close for comfort. Romney's Mercury and Sun also fall in the election 2nd. It's a clear message: Romney wants our money.

Romney's Moon/Jupiter, which is obscurely placed in Romney's chart, is in the 10th of SC primary day. When Mitt said he enjoyed firing people, he was heard loud and clear.

Note that Venus and Jupiter are in mutual reception, second to third. Mutual receptions push the houses the planets are in together and would ordinarily make for heavy turnout (we value (2) voting, so let's go (3) vote!). In this case, as Jupiter is trapped late in the third, the expected heavy turnout does not happen.

Now Newt. His Sun is in the opening's 6th house. He looks like one of us (a worker bee). His moon is in the 12th, where it secretly understands, and where it also hides his cruder side. His Jupiter is setting, which gives the people hope. His Mercury is at the opening's 5th house, where it appears innocent and childlike. His Mars is in the 3rd house, which it rules. Is Newt the savior? Of course not. His Pluto falls, intercepted, in the 7th house, which means he will betray the voters. His Neptune is late in the opening's tenth, meaning he is being dishonest.

The polls will close at 7:00 pm on Saturday. **The close of the polls tells us the outcome. It does not tell us *who* will win, but rather, *what* will be won.**

At 7:00 pm on Saturday, 18 Leo rises. Sun, the ruler, is debilitated in Aquarius. Which means the results will not favor the winner, whoever he turns out to be. Wouldn't this be true of all South Carolina presidential primaries, since they're always held late in January and Leo will always rise, ruled by the Sun?

No. Not quite. The last time, January 19, 2008, the Sun was in the last degree of Capricorn. Capricorn is the sign of the government, which

Newt Gingrich Hates Mitt Romney 55

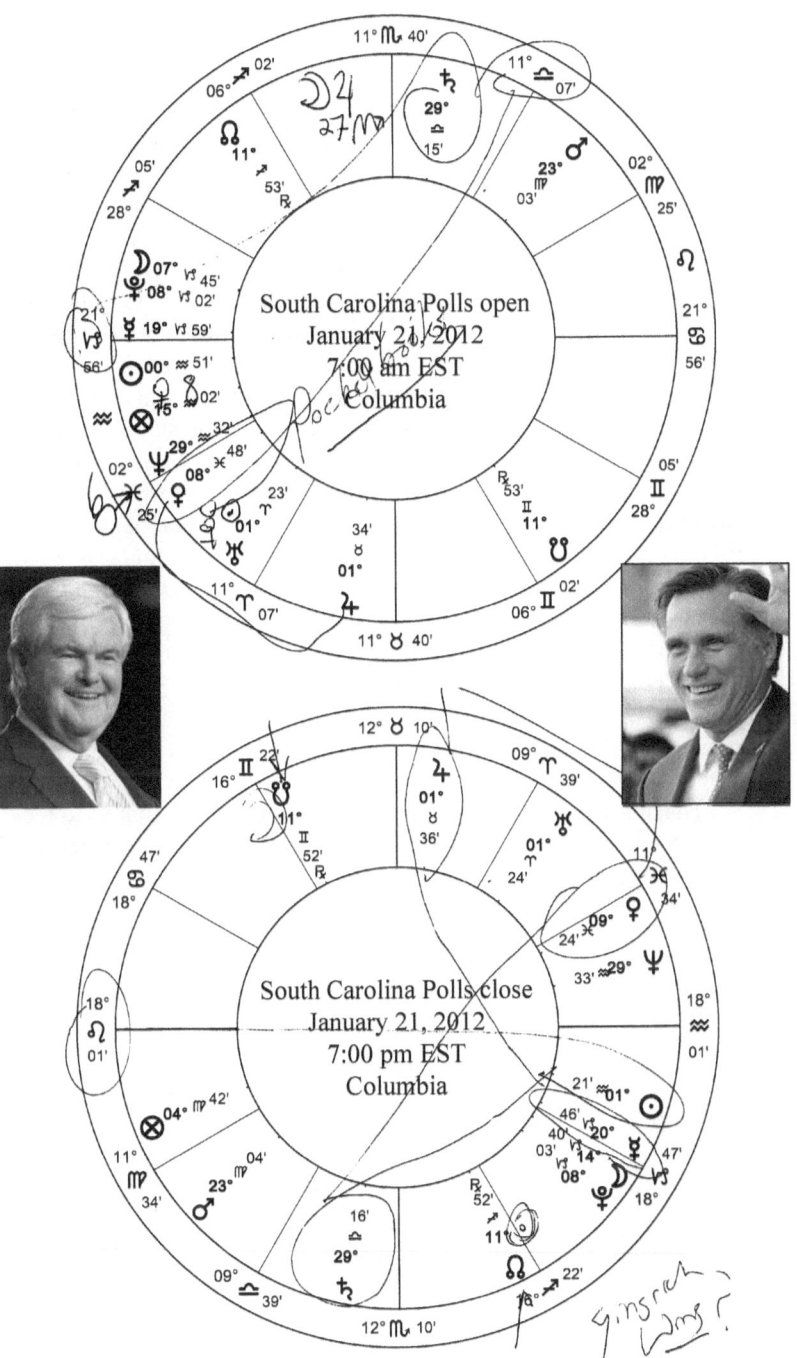

explains the tradition that whoever wins South Carolina goes onto win the nomination. Not so this time.

Sun in the 6th in a mundane chart represents labor and, in a larger sense, business. Debilitated, it means the election was not good for working people. ("I like firing people." — Romney.) Debilitated Sun ruled by Saturn, in the third, the bad news is known to all. Saturn in his turn is ruled by Venus, who is now, in the closing, on the 8th house cusp of other people's money. Whoever wins the primary will get lots of money as a result. This money will come from Venus's ruler, which is Jupiter. As before, Jupiter and Venus are in mutual reception, and, as before, Jupiter is trapped by house. The money the winner hopes will continue his campaign will not come from religion, since religions will support openly. The money will come from foreign governments, whom we may presume are hoping to buy a government that will not make war on them. Or, in the case of Israel, a government who will remain obedient to foreign masters. Such is the cruel stupidity of *Citizens United.* If corporations can out spend ordinary citizens, governments can out spend corporations.

Mitt is a master of other people's money. His natal Sun and Mercury are in the 8th of the election outcome. His Moon-Jupiter are "at home" in the 4th. So, does Mitt win?

My prediction: Mit loses. Newt wins. I write on the Sunday before election day. Already Gingrich has pulled even, but he will not be the nominee in the fall. This is not to say I like him, or anyone else who is in the race.

The closing chart has Newt's ascendant/descendant, almost to the exact degree, but reversed. Newt's full moon falls in the reverse houses from his natal, with the nodes nearby.

Election 2012 will henceforth be about money and jobs. All we have to do now is find someone, anyone, to vote for. — *January 17, 2012*

Newt wins South Carolina

Do not underestimate the power of a full moon with a mission (Gemini-Sag). This is Newt's last hurrah and he knows it. Saturday's win has put a taste of victory in the man's mouth. He will be the next president or will destroy the Republican party if he is denied. There was record turnout on Saturday, so I got that one wrong. And the daffodils, which were 2½ inches, are still 2½ inches, only now with ice. — *January 24, 2012*

Hellenists
Medievalists

With the arrival of Ben Dykes' *The Book of the Nine Judges*, and Chris Brennan's article on Hellenistic astrology in the new issue of the *Mountain Astrologer*, I'm starting to wonder if the revivalist movement has not peaked. Prof. Dykes has given us a translation of a translation that was not very well done the first time, and Ben regrettably did not take the opportunity to improve things. Mr. Brennan has given us a nice summary of Hellenistic astrology. With both Dykes' medieval texts, and Brennan's Hellenistic approach, we are asked to learn a great many techniques but the result is only a slight improvement on what we could do twenty years ago.

This is in large part because the people at the forefront of these movements are either traditional psychological astrologers (either directly, or their teachers) or are university professors rather than actual astrologers. So, on the one hand, Chris Brennan has a largely aspect-dominated astrology (which seems to date from Charles Carter's 1930 masterpiece, *The Astrological Aspects*), on the other, Benjamin Dykes gives us a strictly academic approach.

Properly used, houses, signs, planets, dispositors and rulers produce such an amazing and powerful amount of detail that I cannot understand why I would need to fuss with either the medieval or the Hellenistic approach. Virtually every serious book published in the last 300 years has the essence of this system already laid out, but the books that teach how to use it are few: Patti Tobin Brittian's *Planetary Powers* is brilliant. Hamaker-Zondag's *House Connection* is okay, if dated. Morin's *Book 21* will introduce you to his system. I should be lecturing at UAC, but that will not happen, as there are somethings my full moon cannot do. — *January 24, 2012*

⏰ Whole Sign Houses

There were a number of requests to expand on my notes last week on whole sign houses. On Friday the *Mountain Astrologer* for February/March arrived and in it I found an article by the inestimable Chris Brennan on Hellenistic Astrology, including notes on the Hellenistic use of whole sign houses.

The *primary difference* between my approach, and that of Chris, is that he has read and studied and seeks to apply what he has learned. Whereas I skim books and invariably fall asleep in class. To compensate, I make things up and then try them out to see if they work. There are disadvantages to both of these systems, as we will quickly learn.

"Whole sign houses" is a Hellenistic term and we will therefore start with the Hellenistic definition: We take the degree of the ascendant and apply it to all the other house cusps. Isn't this the same as Equal houses, you ask? Not exactly. Equal houses start with the degree on the cusp, and continue to the degree of the next cusp.

Whole sign houses start with 0° and continue to 30°. The degree on the cusp is merely an interesting artifact. The exact degree on the ascendant is a sensitive point, but that seems to be about as much as the Hellenists mess with cusps *per se*.

Since the ascendant is going to be important, the first thing we need to do is *not* grab our existing ascending degree and slap it 'round the chart, but instead use Hellenistic techniques to find out what our proper ascendant should be. And while the Hellenistic ascendant should be the same as the modern ascendant, I don't think we should overlook this step. If the two do not match, we have an obvious problem.

I am not in possession of a comprehensive survey of Hellenistic methods for determining the ascendant, but I do have my nearly finished version of Valens to hand. From the *First Book*:

Book 1, Chapter 4, Valens says to take the Sun's degree position, note where its dodekatemorion falls. The sign in trine to the left will be

Whole Sign Houses 59

the Ascendant. My Sun is at 21♒ and I am a day birth. The Dodecatemoria is 3 ♏. This is actually quite close to Valens' own example, Sun at 22♒. Therefore a day birth will have Pisces, or Taurus or Cancer as the ascendant. There is no mention of an exact degree ascending.

Second method: But if you don't like the first method, Valens has others. His next method is to take the Sun's degree position and add to it the rising time of the sign the Sun is in. Rising times are essentially Alcabitius houses. The system was invented, not by the Arabs, as you might have been told, but by Hypsicles (ca 190 BC - ca 120 BC) who divided the day into 360 equal parts, where each part was equal to four minutes of clock time. Why does Valens not use this as a house system? Presumably because he had not a number system that would let him. The various Greek number systems were little more than hash marks. Roman numbers were already an improvement. But I have digressed.

Anyway, my rising time for Aquarius at klima 4 (my latitude of birth, expressed in Ptolemaic notation) is 22:00. (These are four minute units. The actual rising time is 88 minutes.) Add 21 for the Sun to get 43. Count off from the Moon in Leo and I get Aquarius rising. Which is to say, a sunrise birth. But no exact degree.

Method no. 3, find the number of days from August 30, the Egyptian New Year, to the day of birth (never mind what that number is). Multiply the hour of birth (7, in my case, counting from sunrise, not midnight) by 15 and add this to the number of days. Count from Virgo, giving 30 to each sign. This results in Taurus.

Fourth method, for the "mystical, compelling ascendant": Multiply the hour of birth by 15 (7 x 15 = 105), add the degree of the Sun (21) to get (126). Divide this by the rising time of the Sun's sign at the klima of birth (22), count the result from the Sun's sign to get 5.75 signs from Aquarius. Which is late Cancer rising. Conception will have occurred in the hour of the sign opposite, which in my case would be Capricorn.

Fifth method, to get an exact degree: Multiply the hour of birth by the motion of the Moon, and then for a day birth, count from the Sun's position. On the day of my birth, the Moon travelled 11°56', which is to say 12 degrees. Twelve times 7, the hour of my birth, gives 84. 84 + 21♒ gives 12♉ rising. At last, a number!

Sixth method, for day births, such as mine: Add the remaining degrees in the Sun's sign (9) to the Moon's position (139), then divide by 30. The remainder (remember remainders?) will be the degree rising. Which in my case is 28. Presumably of Taurus, since that keeps coming up.

Seventh And Final Method: Count from July 19 (Epiphi 25) to the day of birth, then add 22. Starting at Cancer (for a day birth), count off by 30. The ascendant will be where the count stops. The remainder will be the degree rising. This gives me 18♑ rising.

Humble reader, **I in fact have 14 degrees of Gemini rising.** A sign that Valens never once produced.

Caveat: Valens talks of sundials and the gnomon of the sundial. The gnomon is the upright part that casts a shadow. One easy technique that Valens did not give, perhaps because there was no need, was, for day births, to note the hour of birth and then read off the ascendant directly from a weekly table of shadows.

Yes. A common sundial and a simple table will give the ascending degree for a day birth without any calculation whatever. Which means the great Egyptian obelisks were in fact sundials. As their shadows were very long, they would have been extremely precise timekeepers and should have easily given the ascendant to the exact degree. Sunny climate, too. This would have been invaluable for horary work, but again I digress. Open at dawn, close at sunset.

Despite Valens' failure to find my correct ascending sign (never mind the degree), his method for finding the midheaven was much simpler.

Add the rising times from the sign on the descendant, to the sign on the ascendant, and divide by two. Since the consensus is that I have Taurus rising, if I give myself a putative 15♉ rising, then at klima 4 (38 N latitude), the rising time from the 7th house cusp, 15♏, to the ascendant at 15♉ is 145:20. Half of this is 72:40. Adding 72:40 to 15♏, (215), I get 23♑40 as the MC. Using the computer, I get 28♑. Using this same method with my actual ascendant, 14♊, I get 23♒40. My actual MC is 22♒. Which, given the amount of fudging involved (klima 4 runs from 36 to 41°) the Valens method of calculating the midheaven (he only gives one method) is NOT BAD. I was impressed!

It was for this reason I said that Andrea Gehrz's translation of the first book of Valens was what you were expecting. Modern Hellenists don't think ancient calculations are important and so have ignored them. In fact, they are critical. If the Greeks could not actually determine an ascendant, which I think to be the case, then the delineation and forecasting methods based on it will be whimsical at best. From this and other study of Valens, I have concluded that Hellenistic astrology was in fact a work in progress. Not a perfected system. Which, by the way, was the medieval opinion. To medieval astrologers, the Greeks were an open book. Not a new and exciting discovery, which the Greeks are to us.

At a later point in Valens (I've not yet indexed the book, so cannot quite find it), he proposes trisecting the arc between MC and ascendant. Which are Porphyry houses. Pure and simple they are.

For his part, Brennan does not mention the midheaven, presumably because he thinks the Hellenists did not use it. In his TMA article he mention Valens by name, so the omission of the Valens' MC is curious.

But as I say, I'm too lazy to stay awake in class, so I have to go about reinventing the wheel for my own sake.

Hellenists and Westerners in general have always said the Indian civilization was borrowed from the west. India, for its part, has always disagreed, admitting that while they do indeed compulsively borrow, their civilization is their own. In the analysis of Hellenistic methods of calculation we have testimony to support the Indians. Traditional Hellenistic calculations, as given by Valens, do not work, whereas, traditional Vedic calculations do. So who copied who? Charts that lack midheavens are valid in equatorial climes, such as that of India, where the MC is rarely more than a few degrees from exactly square to the ascendant. But this is not the case at Alexandria (32°N), where the MC can wander quite a lot from where it should be. The problem is even worse in the Crimea (Sebastopol: 44°N) where I believe Vettius Valens actually lived and worked. Which is why he, and northern Europeans in general, found midheavens to be important, and why the latter invented so many ways of dividing the arc between MC and ascendant.

Forgive me, but I will now set modern Hellenistic fantasies aside and show you some interesting things about houses, as I use them.

I use Placidus houses because, well, everyone uses Placidus. Soon after I started studying astrology in the mid-1980's, I was stuck with the charts of friends who had no birth time. In frustration I would try to guess the houses where I thought their planets would be. I would spend several hours fussing about, this way and that, and would then go to my tables of houses (the *Rosicrucian*, in fact), turn to the pages featuring the latitude of birth, and find the line(s) that put the planets in the houses where I wanted them. This always worked. I did not think anything about it.

At the time I worked for the New York Astrology Center and at the time everyone was crazy for Koch houses. Which had been invented a dozen years earlier by Dr. Walter Koch, of Germany. He claimed his system was the only "true birthplace" system, which meant that if you weren't using Koch, you weren't getting German precision, and we all know how precise the Germans can be.

So one fine day I had put all the planets in the houses where I wanted them to be, had carefully noted their zodiacal positions, and then, for the first time, turned to a Koch table of houses to find the "real" cusps.

I looked and looked, but to my surprise, there was no entry, there was no line, that put the planets where I wanted them. It was not there. It was at that moment that I realized that houses were not trivial affairs, that, whatever they were, there was "real stuff" going on with them. I closed the Koch book, I have not been back since.

Koch cusps in fact are sensitive to common transits. The intermediate cusps, that is. Koch angles are exactly the same as Placidus. Indeed, virtually all quadrant systems (Regiomontanus, Campanus, Topocentric, Porphyry, Placidus, Koch, etc.) give exactly the same angles. All the fuss is about the 2nd, 3rd, 5th, 6th, 8th, 9th, 11th and 12th cusps.

So the first clue was that Koch could not be used for delineation, only transits. As transits to house cusps were trivial, I set Koch aside. Placidus cusps had value.

The next question had to do with planets which were not in the same sign as that on the cusp itself. Working this out led me to whole sign houses (the topic under discussion this week), but there are a number of kinks.

First, we have to realize that signs are real things. Not wispy, vague or ephemeral. It is commonly held that a planet 5 degrees outside a house is to be considered as actually in the house, but it seemed to me that if the signs did not match, the planet could not "jump into" a house of a different sign.

Which automatically meant that a planet in the same sign as the cusp, but further outside than 5°, was probably in that house as well. This problem, of a planet outside a house, in a sign that was, or was not, the same as the cusp itself, could only be resolved by whole sign houses. All planets in the same sign were in the house with that sign on the cusp. Sound simple? It isn't.

What about 28 degrees on, say, the 5th house cusp, with Mercury at 4 degrees of the next sign? With our rectification system we've already established the Placidus cusp as being the right one (Koch is never that far off, but it clearly does not work), so that means Mercury is in the 6th?

But the actual 6th house cusp, at, say, 25 degrees, is a long way away. And in these cases, the native does not act as if his Mercury is in the 6th. He acts very much as if his Mercury really was in the 5th, even if whole sign houses said no.

So I made a fudge. I said, look, the house is like a 100 yard sprint. At the start line (the cusp) there is the blue team and things to do. At the finish (the next house cusp) there is the red team. For whatever reason you, a red team member, have found yourself standing on the blue 10 yard line. Do you run, run, run a long way to that distant cusp, or do you wander over to the blue team?

Before you insist on purity, that a house can be mostly in front of the cusp, rather than behind it, based entirely on the Placidus cusp, consider the angles.

Whole Sign Houses 63

As it happens, I was born with Sun opposite Moon, Moon conjunct Pluto, Pluto tightly opposed to the Sun. When my chart was first calculated I was delighted to see that my Sun and Moon were in fact smack on the MC/IC axis, a rare thing and obviously Important and Powerful. I felt privileged.

But I could not make the delineations work. Sun in 10, I'm supposed to shine, but, actually, I don't. Well, maybe Sun in Aquarius in 10 wants to defer to others and make them shine. Wouldn't that make me an ideal master of ceremonies, introducing others and letting them shine? Well, yes, except that, well, I don't. Go read the poisonous reviews I write.

Moon in 4, conjunct Pluto as well, I'm going to kick my family's ass and make something of them. Family is important! Family is essential! And by the way my old man (4th house is daddy) was a Terror!

Except that none of that was true. My father and I were never close. I have four brothers and four sisters, all of whom are still alive, only one of whom I am in contact with. So far as my own family is concerned, I was 48 before the birth of my first (and only) child, and 56 before I finally got married. Property? My first mortgage came in time for my first child. That isn't 4th house.

But then I suddenly remembered the little mental trick I taught myself in the 8th grade: That I could put questions to myself, wait, and suddenly there would be an answer. That when I was a teenager, I could argue logic in completely contradictory directions. That, at the age of 18, in frustration, I gave up logic and went to stream of consciousness. Which, today, is all I know. As a result I have long been of the opinion that I can know anything merely by thinking about it.

And finally it came to me: I had a Sun-Moon polarity, not in 10 and 4, but 9 and 3. I looked again at my birth time, 12:32 pm. What if there was a transcription error? What if the doctor wrote "12:52", but the clerk who transcribed his notes and filed the birth certificate wrote "12:32" instead? Mistook a 5 for a 3? I can't tell my wife's 5's from her 3's.

So I advanced the chart by 20 minutes and to my astonishment, I found the Sun, Moon and Pluto had all nudged back, ever so slightly, into 9 and 3.

Ever so slightly. Excitedly, I ran progressions and for the first time I got a real result. I quickly established a birth time of 12:47 pm.

What are the numbers? Correcting for a simple 20 minute transcription error, at 12:52 pm the MC is 23♒28. Moving backwards 5 minutes gives a final rectified MC of 22♒11 at 12:47 pm. The Sun is 20♒59. One degree, 12 minutes shy of exactly conjunct the MC. Yet the strongest planet in the chart, at the second strongest point, do not connect.

Nothing excites me more than puzzling out something entirely new.

I have set myself challenges, such as finding the cause of Mozart's death, or how astrology works, and found innovative and novel solutions. This is THREE and NINE. NINE and THREE.

Four and ten, I would be at UAC in May, I would promote myself ruthlessly, but as I have not a trace of Four and Ten in me, I will never be there. In fact I will never be asked, for I am a 9th house ideologue who has burned his 10th house britches. Believe me, if my Sun was in 10, if my Moon were in 4, burnt britches could not stop me. In puzzling out my own chart, I was shocked to learn the full strength of the Sun and Moon.

I have no reason to think planets just shy of the ascendant would not produce the same results. Which makes angles unique. It makes them walls.

So our whole sign house theory is evolving. Sometimes a whole sign is the whole house, sometimes it's not. It depends. You have to look at the chart and deduce.

Cadent houses (3, 6, 9, 12) are different from the others. They are larger, for one thing, as they start several degrees in advance of their cusps, and extent right up to the angle that follows.

Cadent houses customarily respond to not one, but to two different signs. If both signs are empty of planets, the sign on the cusp rules, by means of the sign and house of planet that owns it. If there are planets in the same sign as the cusp, then traditional dispositor rules apply.

If there are planets in the "left over" section of the next sign, those planets, by definition, are stranded. They are like a "surprise." I am of the opinion that such planets act out. Here I am thinking of Michel Gauquelin's work. From 30 years ago, he discovered great planetary strength, not in the 1st and 10th, which would be expected, but in the 12th and 9, where it was not expected. With few exceptions, planets in the Gauquelin sectors would be in the same sign as those of the angles, the ascendant and MC. Stranded planets act out, encouraged, I suspect, by the fact that, although there is no "angular house" backside, there is a nonspecific energy that powers them.

We look for analogies to angles and find them in the obelisk. The ascendant is the moment before shadows begin. The descendant is the last moment that has shadows. At the midheaven, first we have shadows on one side of the obelisk, and then in the next instant, we have shadows on the other. The transition, from no light, to light, from light to no light, from sun on one side, to sun on the other, is instantaneous. The angles are times and places of transformation, of magic itself.

There is one further detail: Intercepted signs. Crudely speaking, interceptions happen to those born north (or south) of the tropics, when the midheaven and ascendant are far from 90 degrees apart. When they are 70 or 130, two of the four quadrants are huge, the other two are tiny and in the huge ones we will find signs entirely swallowed up in one house or another.

Going a long ways back in this essay, if a house cusp illuminates and defines the sign which is on that cusp, then a sign that lacks a house cusp has no direct way of expressing itself in the chart. It is stranded, like a ship without a rudder, or without a mast, or without an anchor. Such a sign is by definition excluded from directly expressing any of the twelve cusps. If this sign is empty there is no harm.

If the intercepted sign has planets in it, those planets are essentially trapped. They will be of the house they are in, but they will express it poorly. The house will be known primarily by the sign on the cusp and the sign and house of its ruler. Not the planets trapped inside it.

With planets in the intercepted sign, there will, at the same time, be an entirely different way, or method, of dealing with the affairs of the house. The relationship between these two methods will be shown by the relationship between the two rulers, and if there is no relationship, no aspect, if one does not rule the other, then the two methods will exist independently of each other. The native is often completely unaware.

When there are planets "in the forecourt," which is to say, in the last few degrees of the sign on the cusp, those planets, and the ruler of the cusp, dominate the house, the intercepted sign is ignored.

When both the forecourt, and the intercepted sign, both have planets in them, the resulting relationships can be complex. In my opinion, the planets in forecourt have the initial advantage, but over time the planets trapped inside the interception will come to dominate. This will produce problems.

Such is a sketch of "whole sign houses." While the overall concept is simple, the details can be complex. Your guide is not a set of rules to be memorized and recited, but your own head. Increasingly I find astrology to be brutally literal. If you have a planet in a house but not in the same sign as on the cusp, and if that planet is rather close to the next cusp, and that cusp is not an angle, then that planet is in the next house. Not the one you think it's in. As I've been reading charts for astrologers, a number of them have strongly disagreed, but the proof is in the delineation.

Which side of the cusp a planet is on is important. Which is to say that those who use Placidus houses to rectify not only put planets *in* houses, but also exclude planets *from* houses. Such excluded planets would presumably be "loners."

Planets in a house, in the same sign as the cusp of the house, are therefore comfortable in that house. Whereas, planets that were in the same sign as the cusp, but which fell outside, were "struggling" to get into the house, or "running to get into" the house, etc. It is also clear that this "running" had to do with the planet's direct motion. A *retrograde* planet falling *outside* the cusp of the house it was associated with, would obviously *not want to go there*. He would be reluctant. He would *shirk his duty*. There are lots of details like this.

Which is why you should have your chart read by someone other than yourself, someone hopefully who is competent and can actually read a chart. You gain a useful perspective on yourself that way.

Frankly, I have seen some really sad charts. Next week I may bring you the chart of a reincarnate female slave. In her last life she was forced to breed against her will. Slavery in America, which lasted some 200 years, easily outstrips the Nazis in terms of sheer horror, as the chart I have in front of me mutely documents. Unlike the Nazis, who came and were quickly gone, the two centuries of unending terror of American slavery is still far from finished. The mass rape of females, the mass castration of males, the forced breeding that was slavery in America, was one of the greatest crimes ever committed on this Earth. It will take centuries to heal, as its victims reincarnate with nightmarish charts, for which they are not responsible, and over which they have little control, for that is the very definition of slavery, that you have no free will. Modern Americans have a unique opportunity to learn and be humble.

In this case, Saturn-Venus conjunct in Virgo, intercepted in the 5th are the sickly children she was forced to breed and which she does not admit to be hers, since Leo on the 5th house cusp will not "see" them. The Moon in Pisces intercepted in the 11th house, she disowns them, but, as Venus is debilitated, Venus in the 5th, it wants to run across the chart to Pisces to get the Moon's (mommy's) attention. Which, literally, is what children compulsively do: Search for their mother. So far as the individual herself, who asked for my help but who has not responded, it took some effort to arrive at a simple solution: She wants to be left alone to do her work. With all my heart, I wish her peace. — *January 24, 2012*

⊕ The Florida Primary

The Florida primary is on Tuesday, the polls open at 7 am EST. Most of the state is on Eastern time, but the panhandle is Central, so polls there will open an hour later than they will in the peninsula. They will also close an hour later, at 7 pm **CST**.

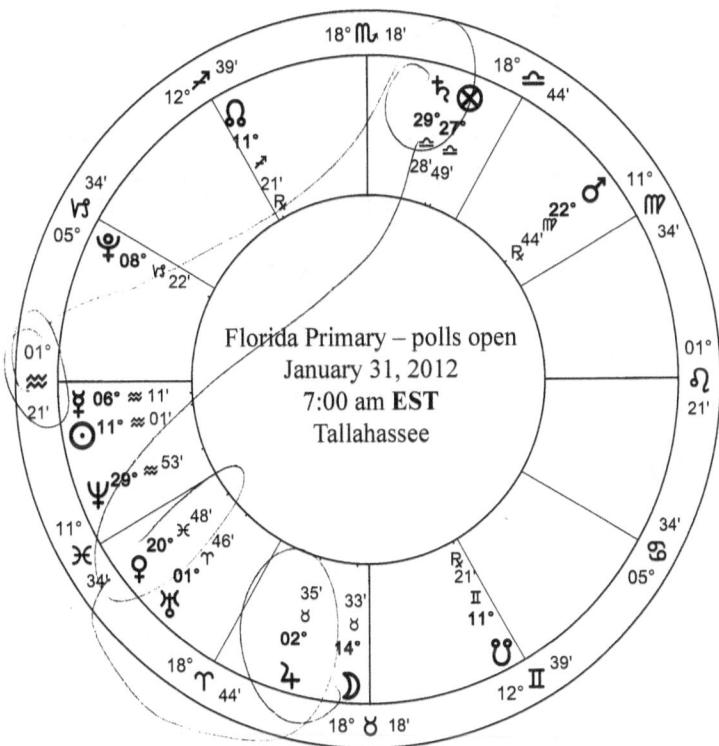

What will Floridians be voting for? Or against, as the case may be? Set the chart for the state capital, Tallahassee, for 7 am. One degree

67

Aquarius rises, the ruler is Saturn at 29 Libra. Of Saturn at that degree, Helen Adams Garrett says, *"Very much or very little cultural atmosphere, music, clothing, jewelry, art, etc. Is learning through relationships not to be possessive nor to allow abuse."* Which I think means it's been a cheap and nasty campaign. Parlay this into the 9th house and people will be voting their religious beliefs, which will be very close to bigoted. Which would be bad for Romney (Mormons get the Mormon rap) but not necessarily good for anybody else. (Those of you reading in future years, yes it really was that bad in 2012.)

What kind of beliefs? Look to Saturn's ruler, Venus. It's in Pisces in the second: Pocketbook. Since both Venus and Saturn are strongly placed by sign, voters will be voting *for* someone, rather than against. For someone who has promised them money, if not jobs. Newt had the most outrageous proposal, to establish a permanent manned base on the moon by the end of the decade, which will be manna to the ears over at the Cape. Venus in Pisces, it went over quite well. Which, crudely speaking, are two points for Newt.

On the day Venus will be conjunct Romney's Sun. She is smiling at him. The ascendant is exactly trine Romney's. Romney wins, but you already knew that.

What will he win? Run a chart for 8:00 pm EST for Tallahassee, the closing of the polls. Oddly enough, Romney wins Gingrich. It's Gingrich's Mercury that's exactly conjunct the MC, so look for the blow-hard to dominate the evening's election returns. Romney's Mars is on the 7th. That sounds like buyer's remorse. Ron Paul does surprisingly well. His Moon is conjunct the Moon of the day, which makes him popular. His Venus is exactly conjunct the primary's Mars, but both retrograde means he has the gays. There is a Gays for Paul website.

The longer the primaries go on, the worse it gets. — *January 31, 2012*

☾ Madonna

Madonna, Queen of Super Bowl 46 (February 5, 2012), was born on August 16, 1958, at 7:05 am, in Bay City, Michigan. Her chart is not all that hard to read.

Virgo rising, ruling planet Mercury in Virgo, Moon in Virgo, Pluto in Virgo, Madonna Louise Ciccone is exactly what she appears to be: Obsessed with details, to the exclusion of most everything else. All four angles mutable, her life is a chaotic whirlwind, ever changing, never really under her control.

The four mutable signs have but two rulers: Jupiter and Mercury. Interestingly enough, these four signs, Gemini, Virgo, Sagittarius and Pisces, have the same two planets in detriment: Jupiter and Mercury. As such, in charts with mutables on the angles, there is a great emphasis on these two planets alone. Mercury at 5 Virgo, retrograde, and Jupiter at 26 Libra, their angular separation is 51 degrees. Which is a septile, a weak aspect.

In Madonna's chart, Mercury rules the ascendant, Virgo, and the midheaven, Gemini. It is powerfully placed in Virgo and very near the ascendant itself. Jupiter rules the descendant, Pisces, and the 4th house, Sagittarius. We find Jupiter obscurely placed at the end of Libra and at the very end of the 2nd house, with 0° Scorpio, not Libra, on the 3rd house cusp. Jupiter is not only remotely placed by house, he is also on the weak end of a weak aspect, because Mercury, reinforced in its own sign and very near the ascendant, is senior in the septile. Jupiter, conjunct the north node at the end of an otherwise empty second house, is stranded.

Given that a weakly placed planet will not rule its signs very well, we find that instead of Jupiter ruling the 4th and 7th from strength, Mercury in Virgo rules, but in a negative way. Mercury is debilitated in Pisces and Sagittarius, which accounts, in part, for Madonna's failures in marriage (7th house), as well as her inability to stay in one place very long (4th house).

In Madonna's chart, Mercury is in the 12th, three degrees from the ascendant, and in a sign that it rules. It is retrograde and so, "walking away" from the angle. I've made much of the angles being "walls," so are Madonna's Mercury and Pluto in the 1st or 12th?

While I could be sloppy and say of course Madonna's Mercury and Pluto are in the 1st, it is not clear that they are. Madonna has never been known for her ability to communicate. Not even, Mercury retrograde, communicate thoughtfully. Right from the start, Madonna was about spectacle, Madonna was about shock.

Thinking about this carefully, shock is what we would get if we took 12th house secrets and smeared them across the ascendant. The 12th house is like what's in the bottom of the drain in your kitchen sink. The sink looks so innocent, it looks so clean and sweet, so much like, I dunno, a *virgin,* but when it clogs and stops up, you get a plunger and out the junk comes. And it *stinks!* But you never knew.

In Madonna's case, Pluto in the first degree of Virgo (making her of the Virgo generation), in the 12th house draws up raw, nasty, disgusting, repulsive things. Pluto in Virgo, Mercury owns him, and as Mercury is also in Virgo, Mercury takes on everything Pluto churns up. Mercury also owning the ascendant, he can dump it there for all the world to see.

The 12th house is the house of secrets, and while Madonna has Leo on the cusp, Pluto in Virgo is working a different mine, one full of nasty details. The 12th house has to do with the past, and also about institutions. In other words, institutions with a secret past. There is no bigger, no older, no more secretive institution on the planet than the Church. You will say, isn't the Church the 9th house? Yes, so far as beliefs and rituals are concerned, it is.

But so far as a bureaucracy, made up of a vast group of people all under one command, the Church is 12th. Astrology is neither simplistic nor monolithic. Belief in the Trinity of God is 9th house. Go to church every Sunday and do what the preacher tells you, be a cog in the wheel, and you are in the 12th. Astrological symbolism changes when your point of view changes. When your question changes.

Add Virgo to the mix and you will get sexual hangups. Do you know why religions, which is to say, the *9th house,* despise sex? Because religion is about liberation of souls *from* the earth, whereas sex is a *mechanical function* that brings souls *to* the earth. Sex is a one-way street that *brings you here.* Religion is a one-way street that *takes you away* from here. Put more crudely still, sex is the "religion" of the young. Religion is the "sex" of the elderly.

But long ago people forgot that. Instead, power-mad clerics imposed one rule for all: Sex is bad. Bad for you, bad for me, bad for the young, bad for the old. Bad, bad, bad. In other words, a logical philosophy (sex goes this way, religion goes that way) got taken out of where it belongs — the 9th — and was transported to a bureaucratic, repressive, imprisoning 12th. Where it remains to this day. Madonna's chart makes her uniquely qualified to expose this fraud, to bring all its gory details to light. Which she has done. No surprise the Church screamed in agony.

If Madonna's Mercury was actually on the ascendant, she would be known for communication, for intelligence, for detail, for precision, for purely Mercurial traits. Instead she is known for Pluto, and for Pluto alone, which means Mercury's role has been reduced to that of a mere messenger, a go-between, between Pluto and the ascendant. This can only happen if Mercury is in fact in the 12th.

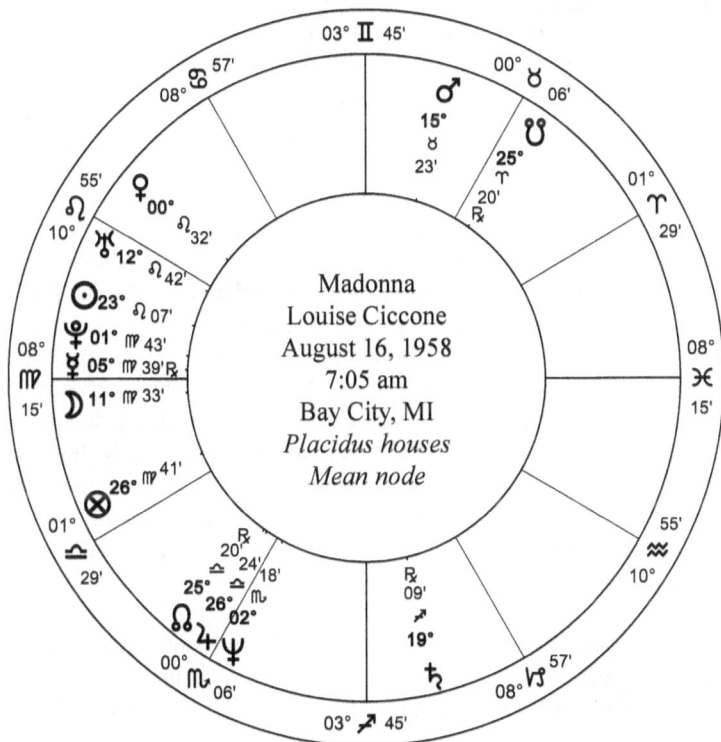

Mercury's retrograde condition reinforces it. Madonna's retrograde Mercury repeats itself. Which is a Gemini trait. In Madonna's case, she repeats the same shocks.

On the other side of the ascendant, in the first house proper, the Moon in Virgo looks on with indifference. What have you brought me? Oh. I see. Clean it up and make nice.

For its part, the Sun, in the 12th but in Leo, knows nothing of this

Pluto-Virgo mess and so repudiates it. Result: Madonna's own statements that she has just been horsing around. Not promoting promiscuity. Not blaspheming the Church. That's not Madonna's image of herself. Moon in the first house in Virgo, she thinks she really is sweet and innocent. Conclusion: The ascendant is just as much a wall as the midheaven.

Does a heavy Virgo emphasis sound like a life in music to you? You would be surprised. Opening Carter's *Encyclopaedia of Psychological Astrology* to pgs. 130-1, I read that Virgo is the commonest ascendant for a musician. Carter says that 16 Taurus/Scorpio is important. Madonna has her Mars at 15 Taurus. Carter says musicians have the ends of Leo-Aquarius and the beginnings of Virgo-Pisces prominent. Madonna has her Sun at 23 Leo, Pluto at 1 Virgo, Mercury at 5 Virgo, and the ascendant at 8 Virgo. Her Moon at 11 Virgo is not quite at the beginning of the sign.

Carter says 24 degrees of cardinals is important in musical charts. Madonna has her nodes at 25 of Libra-Aries, and Jupiter at 26 Libra. With Carter's degree positions we must sometimes give him a degree or two of leeway.

A musical life in Virgo will be about detail, precision, and perfection. With Pluto there, with a great deal of raw power and energy. With Mercury retrograde, with thought and contemplation, even if it doesn't appear to be part of the final package. Which means there are no accidents in Madonna's work. Moon in Virgo, there is personal satisfaction in getting the details are right. With Pluto, Mercury and the Moon near the ascendant, with the overall chart ruler among them, when Madonna does gets it right, people respond positively, and when they do, she feels personally rewarded, which is what it means to have the Moon on the ascendant. With all of this split between the 12th and the ascendant, incorporating (with Uranus) a total of six planets, Madonna is essentially a one-trick pony, 12th and 1st.

For Madonna's view of the Church proper, we look at her 9th house. It has 0 degrees of Taurus on the cusp. Taurus is ruled by Venus in Leo in the 12th. Venus says Madonna secretly (12th) likes (Taurus) the Church (9th).

On the other hand, Mars in Taurus in the 9th does not. He prefers a blunt, head-to-head battle. Like a raging bull.

Mars in Taurus trines up with the stellium in Virgo and makes Madonna's attacks blatant and tacky: Mars wrecking the placid beauty of Taurus. The squares to Uranus and the Sun, both in Leo, provoke and energizes Mars further. This is a case of a planet in debility which doesn't want to be in the house opposite. Mars in Taurus in 9 is having too much fun. And it might just be that Madonna Louise Ciccone is a crude woman.

Madonna wants to change the Church, not destroy it. Madonna's conversion to Judaism can be viewed as a rejection of the dead Jesus and his dead religion, and a return to the source, the religion of the living Jesus. If it was good enough for Jesus, then it's good enough for the material girl. Which makes Jesus her goal and Judaism her means. Women inherently sexualize the Church.

So what about sex? Glad you asked. Madonna doesn't get any, or at any rate, she doesn't get very much, and it drives her crazy. What would you think of a person, man or woman, with a Virgo cluster on the ascendant, Capricorn on the 5th and Aries on the 8th? Sexy? No, I don't think so. Repressed? Most likely. So she acts out.

Capricorn on the 5th, the ability to express herself sexually would be delayed, as Saturn slows everything. Saturn in Sagittarius and retrograde, religion would be a reason. To the Church, Saturn represents Satan. Capricorn's co-ruler Mars, in Taurus in the 9th, emphasizes the struggle with religion and sex. Mars in the 9th is also why Madonna would aggressively use mass communications (9th house) in her struggle.

Moving to the 8th house, Aries is on the cusp. Ruled by Mars, this is an aggressive, bold placement that bodes ill for the tranquility of lasting marriage.

Mercury and Pluto account for the loss of her mother at age 5, the most devastating event of her life. Mercury rules the 10th house, which is mother. Using solar arcs, Pluto landed on top of Mercury at age 4. Pluto's subsequent transit of Madonna's Moon signaled the death of her mother.

I am wondering what to do with retrograde planets and solar arcs. Is it possible the arc of a retrograde planet should be run backwards? I.e., converse? If we do this to Mercury, its solar arc ends up on top of natal Pluto at the same time that solar arc Pluto ends up on top of natal Mercury.

Note that when you change the sign on the ascendant, you also change the planet ruling it. Since the overall tenor and quality of the native is determined by the *house* and *sign* of the *ruling planet,* changing the ascendant brings some other planet, some other sign, some other house, into play. If you have a chart where the time is dubious, where the ascendant could be this sign or that one, look to the planets that rule each of those signs, and then look to the houses and the signs they are in.

One house-sign combo will be WHO and WHAT that person is. The other one will be miles off. The choice will be easy.

Madonna is a good example. With Virgo rising, we get Mercury, we get virgin, we get the Church, we get music. We do not get the Sun, since the Sun is hiding in Leo in the 12th.

Change Madonna's time of birth to put the Sun in Leo in the first, and now, instead of a virgin, instead of playing with Church imagery, we get a queen who is proud. Regal. Demanding. Virgo on the second house, the cluster of Pluto-Mercury-Moon will make her a craftsman instead of a sexual extrovert.

It is when you work with planets and signs and houses dynamically that astrology becomes not only easy, but literal as well.

Back in the 1990's, Madonna's life and work were the subject of numerous Ph.D. studies. Might have had something to do with a coffee table book of nude pictures she released about that time. Male academics are addicted to porn just like the rest of us.

Looking at her nude work in my research (I cover all bases), Madonna is one of the least sexy nudes I have ever seen. Moon in Virgo, Mars in Taurus, Capricorn on the 5th, it would be hard to make Madonna a sex symbol. Which might be why academic interest in her eventually faded. For Superbowl 46, she provided halftime entertainment. I've never watched a superbowl as I find football boring, so I missed her show. I presume it was a mixture of well-crafted new material, along with some of the best of her past stuff, and presumably G or PG rated. The gridiron boys are fussy about that. There were no wardrobe failures. — *February 7, 2012*

☽ Lady Gaga

So here is your Valentine: Stefani Joanne Angelina Germanotta, better known as Lady Gaga. She was born on March 28, 1986, 10:55 pm, Yonkers, NY. Let's have a look:

As with Madonna (last week), all four angles are mutable. So, as last week, we look immediately at the rulers, Jupiter and Mercury. We find them widely conjunct in Pisces in the 3rd house.

This is already a very different chart from Madonna's. Sagittarius rises, the ruler is Jupiter which is in its own sign of Pisces. Like Madonna, Lady G's Mercury is retrograde. Unlike Madonna, it's also in debility. In Madonna's chart, Mercury was dominant. In Lady G's, it's Jupiter.

With all four angles mutable, Lady G's life, like that of Madonna's, will be chaotic. Unlike Madonna, who has axes to grind, Lady G's life will be about having more, as that's what Sagittarius and Jupiter are all about: More. More of everything, which is Pisces, and more of being busy, which is the third house.

With Sagittarius rising and with Jupiter as the final dispositor, Lady Gaga's singing style is typically Sagittarian, and here I am indebted to Charles Carter, who, his entry under Handwriting (*Encyclopaedia of Psychological Astrology*), says that, *"frequent usage of the dash is Sagittarian; they often employ no other punctuation."*

Look at this:
Oh-oh-oh-oh-oh-oh-oh-oh-oh-oh-oh-oh!
Caught in a bad romance
Rah-rah-ah-ah-ah-ah!
Roma-roma-mamaa!
Ga-ga-ooh-la-la!
Want your bad romance

Remind you of anyone?
It turned out so right,

For strangers in the night.
Do dody doby do
do doo de la da da da da ya —
"Shoo be doo be doo"

— Frankie, it turns out, had Sun and Mercury both in Sagittarius. While this sounds like scat (a genre of singing), scat is vocal improvisation with its own set of rules. By contrast, making up lyrics because you don't remember or don't want to be bothered or just like the sound of it, is Sagittarian. Sag hates detail.

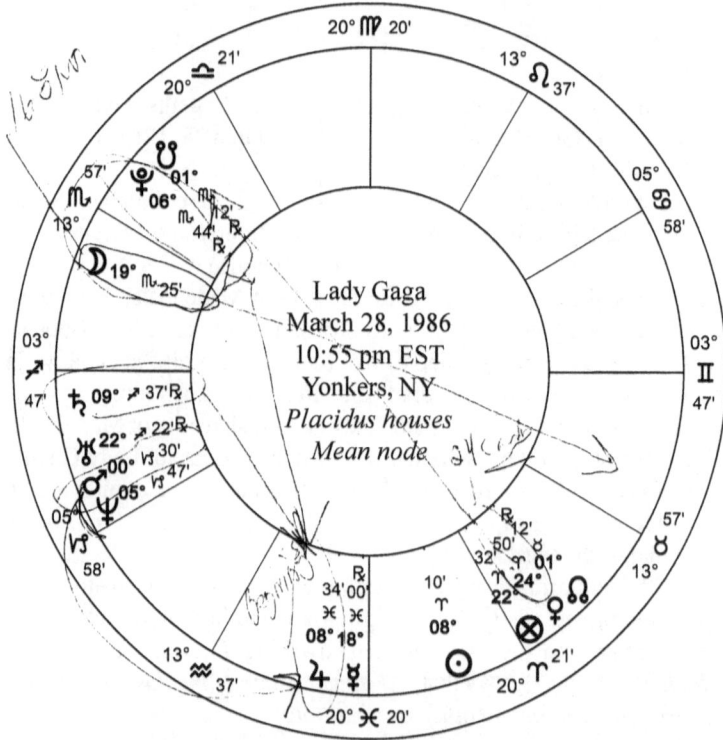

Ruling planet Jupiter is well-supported in Lady G's chart: Trines to Pluto and the Moon, square from Saturn, sextile from Mars. No surprise she was eager to get started in life. She went to middle and high school at the Convent of the Sacred Heart, an all-girl Roman Catholic school in Manhattan. Five years senior, Paris Hilton is a fellow alum. Or would be.

It appears that Lady Gaga did not finish high school. Wiki's exact words are, *"when her time at the Convent of the Sacred Heart came to an end..."* Does her chart show her as a high school dropout? Consider: Aquarius on the 3rd house cusp, ruled by Saturn retrograde in Sagittarius in the first. Aquarius on 3 is *acting out in school* (strike 1), Saturn in the elite sign of Sagittarius would give an interest in the grander side of education, except, retrograde, it won't. In the first house, the disdain is public. *Strike 2*. Jupiter ruling the ascendant from the 3rd house, she's too big for such little things (the 3rd is opposite the 9th, which Jupiter prefers). And that's *Strike 3*.

Did she leave school one day, or was she expelled? Sagittarius and Pisces are mutable (chaotic), Aquarius is fixed. Cardinal is initiative, which is missing. Lady Gaga was expelled.

She got early admission (age 17) to NY's Tisch School of Arts. Which means she applied. Why did she apply? Consider that her 9th house has Leo on the cusp: In college she could be a big shot. The ruler is the Sun, which happens to be in Aries, a cardinal sign (details, details). In the same sign as that on the 5th house cusp, Lady G presumably realized college would enable her to be who she wanted to be: Sun achieving its natural mission in the 5th house by means of its rulership of the 9th. On her Wiki page it says she dropped out the second semester of her sophomore year. Which might be because her Sun is too stressed to sustain the effort, or it might be that Lady G's life is too chaotic overall, or perhaps both.

Further down the Wiki page we learn that, age 19, she was living in an apartment on New York's Rivington's Street (lower East Side, back before it was trendy), which is definitely off-campus. Lady G is proof you don't need a high school diploma or a college education to make it big in the music world.

Judging by what my daughter plays endlessly on the other computer, Stefani Joanne Angelina has had little formal voice training. She has very good pitch. She has recorded, or at any rate, been associated with, several dozen songs. She has made nearly 20 videos and is not yet 26 years old.

It's clear to me that Lady G is running on adrenaline. Can she keep it up? Saturn arrives on her ascendant for Valentine's 2015. The subsequent Saturn return will also square her ruling Jupiter and trine her Sun, but that won't be until Christmas, 2015. By the time she hits 30, in March, 2016, she will be a very different person.

> *And eh, there's nothing else I can say*
> *Eh eh, eh eh*
> *There's nothing else I can say*
> *Eh eh, eh eh*
> *I wish you'd never looked at me that way*
> *Eh eh, eh eh*
> *There's nothing else I can say*

Eh eh, eh eh

Lady G's Sun is in Aries, her Moon is in Scorpio, the aspect is an inconjunct. Ordinarily this would be a case of head not knowing what the heart was grieving over, but this is not the case here. Both Sun and Moon are ruled by Mars, and Mars is in Capricorn, a sign which he himself has rights in. He is newly arrived in the sign, having only gotten there a bare 24 hours earlier. (It is not enough to have a chart. You must have an ephemeris and know how to use it.) Having returned "home," Mars is raring to go, eager to get down to work in Lady G's 2nd house and make the woman lots of money. Via Mars, money is what ties Gaga's Sun and Moon together, and she is more than happy to make the effort required. Martian efforts will be thwarted, however, as Neptune is on the cusp of the 2nd itself.

A Neptune-Mars conjunction says the harder Lady G tries, the stranger the result. On the second house cusp, it means the money she earns will be stolen from her. Can astrology tell us who the thief will be? Of course. Neptune in Capricorn, the planet and sign both ruled by Saturn in Sagittarius in the first house: She will somehow steal it from herself. Bad investments, perhaps. "Somehow" as she will be unaware she has done so, which is because Saturn is retrograde. Did I just hear one of you mutter, "the follies of youth" — ?

Saturn in the first house is hard and severe, cold and unfriendly. As it happens, Saturn is also conjunct the fixed star Antares ("anti-Aries"). Of this, Robson says, *"Materialistic, dishonest through circumstances created by environment, religious hypocrisy, many disappointments, loss through quarrels and legal affairs, trouble through enemies, many failures, hampered by relatives, unfavourable for domestic matters, much sickness to and sorrow from children."* Saturn square to Jupiter indicates financial difficulties. As both planets are strong I would guess the result to be a combination of carelessness (Jupiter) on the one hand, and fatalism (Saturn) on the other, perhaps because of religious beliefs that fail.

That's M-O-N-E-Y, so sexy,
Damn I love the Jag, the jet and the mansion
Oh yea
And I enjoy the gifts and trips to the islands
Oh yea
It's good to live expensive you know it but
My knees get weak, intensive

Sun in Aries wants to be, not in the 4th, but in the 5th. He wants to have fun, do exciting things, be reckless. Wants to have children, which, looking at the rest of Lady G's chart, I think unlikely. Mars in sextile says that if the Sun plans, if he's careful, he can get everything he wants. Why is this important? Because Saturn in square makes it all very hard.

Venus in the 5th would ordinarily give at least one child, except, debilitated in Aries, it would rather be carousing with its friends, in other words, on the other side of the chart in the 11th house. One should never say never, but the only prospect Lady Gaga has for children that I can find is the debilitated Moon in Scorpio. Watch for transits, solar arcs, to it, or to the 8th cusp, or to Mars, which rules it and the 5th. Or to all at once.

Moon. Lady G's Moon is at 19 Scorpio in the 12th. For company it has Pluto at 6 Scorpio, retrograde. The retrograde is why she's not terribly conscious of any Moon-Pluto conjunction, since Pluto is moving away. This is a similar setup to Madonna, who had Pluto ten degrees behind her Moon and whose life was transformed when Pluto passed over it. Lady Gaga's experience was different, in part as her Moon is hidden in her 12th, rather than exposed in the first. According to the ephemeris, Pluto hit Gaga's Moon when she was four years old. Lady Gaga's Wiki page, which was clearly written by the Lady herself, makes no mention of her 4th year, except to say that she took up the piano. The transit might have been what "woke her up."

The classic delineation of a 12th house Moon is they keep their feelings, their emotions, to themselves. Moon in Scorpio, these emotions are intense and dark. Moon ruling the 8th house, it dwells on the experience with the other. It is notable that Lady Gaga keeps her personal life to herself. The Moon debilitated, it's not surprising she considers her romantic life to be a failure, that her men are not "intense enough" to suit her.

Lady Gaga's music is full of cliches and hackneyed lyrics, all having to do with herself. That she is, for the moment, a celebrity, shows how impoverished music and the arts have become.

Which brings up lip syncing. I am surprised this is an issue. People go to "live" concerts to hear the exact recording they bought on-line. They see the singers dance about on-stage while hearing not only note-perfect performances, but the sophisticated arrangements as well, even though many of the supporting players are nowhere to be seen. Like water running uphill, this is an absurdity. We are told that only backup singers are prerecorded, or that the singer's prerecorded vocal track is there "just in case." Which the artists themselves might actually believe, but I assure you that even if the singer really is singing, what you're hearing is a recording, deliberately played back on top of her. Which would be her manager's doing. He has money at stake. He isn't going to put that at risk. She's *untrained.*

It seems that *Sunrise*, an Australian program, accused Lady G of lip syncing back in 2009. Whereupon she forced them to apologize. In searching on-line for video of the offending performance, I found it had been removed.

Contrast this to the Isley Brothers famous song, *Shout*. Shout is two unrelated songs smooshed together with a bit of improvisation to cover the gaps. Shout is what happens when real singers, with real talent, are stuck grinding out concerts and after a while start making things up on the fly. In the case of Shout, the lead singer (I presume Ronald Isley) started with one song and then one night segued into another with the rest of the group following as best they could. Over time the song developed into the show stopper we know today. It is regretted that a live performance of Shout does not seem to exist. In the present sterile musical scene, no such process is possible.

Audience expectation is another factor. Classical performances are never faked, and not because of the difficulties of faking an entire orchestra (which are considerable), but because live performances are superior to canned studio work. I myself thought this unlikely and scoffed, until I spent a couple of years in London listening to BBC Radio 3, which not only features live concert performances every evening, but many live studio performances during the day. The difference between taped and live were night and day, but we are talking about trained musicians.

Looking again at Carter's rules for musicians, we find Lady Gaga has Venus at 24 of cardinal, Jupiter in early Pisces, the Moon near the middle of Scorpio, etc. Not as strong a musical talent as Madonna. Lady Gaga's drive for money is stronger. It's what she wants more of.

I confess I do not expect very much from the arts. Since 1913 we have been in the Age of Aquarius. As of that year all life, all art, all politics, changed from Piscean to Aquarian. I sketched my theory of the Age of Aquarius in the very first issues of this newsletter some five years ago, but I did not do a very good job, so I will give an overview:

Astrological Ages can be interpreted much the same as zodiacal signs, only writ large. They have the same planetary rulers as ordinary signs and will produce much the same effects as ordinary signs, only, as before, writ large.

I learned this in a class at the University of Kansas in 1975, but only came to realize its importance many years later. The class was titled, *Studies in 1913,* and was given only once. I signed up thinking it would be about politics on the eve of World War I. I was surprised to find it was about art instead.

The premise of the class was that "something happened" to art in that exact year. Something in Paris and New York, for the most part. Cubism, for one thing. My mind is simple. I had never understood Cubism, it had looked like gibberish. In that class I learned the secret: Cubism is many different views of the same thing, all presented at once, and all equal to each other. It went along with Dadaism, Futurism, Surrealism and various other trendy things.

It was only much later that I realized the shift in 1913 was from art for art's sake (beauty), to *art as an ideology*. Art that could not be understood merely by experiencing it. Art that had to be explained. *Art that came with a manifesto*. The ideology made it Aquarian. Aquarian meant that Saturn was the ruler, so of course the resulting art was no longer "pretty." The manifesto? Wouldn't that be Mercury, exalted in Aquarius?

But if that was true, it meant that the art of the previous 2000 years, being Piscean, was ruled by Jupiter. Okay. That meant there had been a lot of art and, well, Pisces being mutable, the style changed with every generation. True and true. Pisces and Jupiter both being about religion, there had been a lot of religious art. Again, true.

And then I remembered Venus. This was the very first time I realized that planetary exaltation really meant something. Combine Piscean universality with Jupiter's abundance, a ceaselessly changing style, religious aspiration and sheer Venusian beauty and the result was not only art of great beauty, but art that was, in its highest expression, transcendental by its very nature. You gazed upon it, or stood inside it, or heard the spoken words, or were present in the same room with the players who made the music, and even though there was a man's name attached to the canvas or the page or the score as the person responsible for the work, you simply could not find a human hand behind it. It was beyond human, it was beyond earthly, it was beyond what was possible and one was simply speechless and groped for understanding.

It was Godly, there was no other word for it, but then what if you didn't otherwise need or care for a God and besides how could it be "Godly" since, in every case, a single, known person was responsible and in many cases, the process of creation was actually documented, one way or another? How was it possible that humans could reach such staggering heights?

Perhaps it is that individual humans, on their highest level, are capable of "god" level work. Instead of worship, we should self-develop. Hold ourselves to higher standards.

I then quickly surveyed the other signs of the Zodiac, considering them as Astrological Ages. What was the art like in each of them? Well, in addition to Pisces, Jupiter rules Sagittarius, but as Sagittarius has no religious undertones and as Venus has no special placement, great Sagittarian art was unlikely.

What about Venus? It rules Taurus, so in the Taurean Age there would be pretty art, and indeed we have a few scraps from that age: Fat, earthy things. Attractive, but not staggeringly beautiful. So what about Libra, the other Venusian sign? Pretty again, but, as an air sign, wouldn't it be ephemeral?

I came to a sad conclusion:

The Age of Pisces was, by far, the greatest of all Ages for the arts. To my eternal dismay, it has just slipped through our fingers. It is gone. Vanished like the concluding note of a great symphony. It will be twenty-three thousand years before we see it again.

But since there is art in all Ages and in all cultures, then for the next two thousand years of the Aquarian Age, we will have Aquarian art.

As Aquarius is ruled by Saturn, the art will be bizarre or quirky. Not pretty nor beautiful. Aquarius being a collective sign, this art will relate to some specific group or other. Since Aquarius, as it turns out, is tribal (my group vs: your group), art that is appealing to one group will be rejected by others.

Art that is quirky and bizarre will tend to be confrontational and upsetting. Because of this, art will tend to focus upon individuals acting alone. As the individual artist is the opposite of the collective, the results may be so narrowly focused as to relate only to the artist who creates it.

According to my reduction of an old KU art history class, Aquarius started in Paris in 1913, perhaps with the premiere of Stravinsky's Le Sacre du Printemps, on May 29, 1913 at around 8:30 pm. This, like Picasso's Cubist paintings, was art created by *Pisceans,* which is to say, those who had been born in the last decades of the *Piscean Age.* They expressed their Piscean natures by means of the new *Aquarian* tools that *time itself* had given them.

The next great wave in the arts came in the 1960's. Why was this? Because by the 1960's the last of those born in the Piscean epoch (pre-1913) were retiring and younger men were taking their place. These by definition were Aquarians, which is to say, those *born in 1913 or later.* Because the last of the Pisceans held on so long, an entire generation was skipped. It seems that in music we went from Stravinsky straight into Elvis and the Beatles.

Musically, we went from the large symphonic constructions of 1900, to jazz bands in the 1920's, which were followed by swing bands of the 1930's and '40's, before the emergence of rock quartets of the 1950's and '60's. The process literally went from large groups of otherwise unassociated people (a symphony orchestra: *Piscean)* to quartets composed exclusively of friends *(Aquarian).* There has to be widespread agreement before 60 people will sit down and play a new symphony, but the require-

Lady Gaga

ments for a new rock song are much narrower. Thus did music descend from great, complex abstractions, to mundane courtship stories. A century ago this decline would have been inconceivable.

Since the 1960's music has degenerated further. Rap, for example, relates surprisingly well to 16th century Elizabethan speech patterns. We still have the Piscean concept of art that is lofty and god-given, but what we actually get is closer to self-indulgent reveries. Instead of continual change (a mutable Age), we are now stuck in the "avant garde," a "futuristic" style we can't seem to get beyond. Which is typical of a fixed-sign Age.

Many of you will think I am being harsh, and I am and I must apologize. Piscean art is not so much passé, as it now requires specific training in order to be understood at all. The world has moved on. Now when we visit Chartres Cathedral we see a vast, dark, gloomy place with pretty glass high up in the walls. We are no longer moved, we are no longer brought closer to God or whatever it might be that is greater than our mere selves. Does this sound like small beer? Yes, it does. Stripped of the godly, Aquarius is so much smaller an Age than Pisces.

Meanwhile the study of the great Ages goes on and on. Here is another fragment:

Each Age has their own proper people, those who, by incarnating during that Age, make their greatest progress. Once the Age "opens" (begins), they rush to take birth and will take as many subsequent births as they can before the age "closes" (ends). These are Aquarian Age souls, such as George Bush, Steve Jobs, Bill Gates, Madonna, Lady Gaga, the man known as Barack Obama, etc. Indeed, the vast majority of people now alive are Aquarian souls. While there is nothing that stops souls from other Ages from making an appearance, they will not have rewarding experiences and so will not be eager to return. As you might imagine, in the abrupt transition from Pisces to Aquarius a lot of Piscean souls got caught out, trying for one "extra" life. Upon completion of their present life most of the remaining Pisceans will pack up and not be seen again for many millennia. Such is the way of the world.

Which, if I am not mistaken, will make the remaining Aquarians happier. I wrote this while listening, not to Gaga's music, but to that of Anton Bruckner. Someday I want to study with him. — *February 14, 2012*

◑ *Found:* **George McCormack's Weather Astrology**

George J. McCormack, "Gee Jay" lived from 1887 to 1974. He had a lifetime interest in weather astrology (AstroMeteorology) and was one of the greats in the field. In 1938 he was one of the cofounders of the American Federation of Astrologers.

In 1947 he published *A Textbook of Long Range Weather Forecasting*. James Herschel Holden says it was limited to 100 copies. From what I have in my hands (thanks to Philip Graves of Sweden, who will shortly get that copy), it was 60 single-spaced, one-sided sheets of typing paper inside a three-ring binder. Given that the paper is watermarked and there are erratic typos and strikeouts, it would appear that each of the 100 copies were in fact hand-typed originals. Up until photocopiers, this was what most astrologers had to do to get published at all. And while the content is of priceless value, books such as these, when they turn up as part of someone's estate, are commonly thrown away without a second glance.

I will publish in a month, perhaps sooner, as the book is not big. It will be reset and the typos fixed, just like any other proper book. The price will be $20, perhaps less.

The concepts and detail are astounding. There is no other word that will describe. For the various planetary pairs, there are A, B and C configurations. There are Spring, Summer, Autumn and Winter delineations. McCormack says Arctic air does not sweep thousands of miles south, but that cold air two miles aloft can suddenly descend. His descriptions of planetary effects reinforces my observation, that planets work directly upon the earth, bringing forth energies (e.g. signs of the zodiac) that are *latent in the earth itself.* This is especially true of the noted Taurus/Scorpio earthquake phenomena. The zodiac is not "up in the sky." It is right under our feet. McCormack is a revelation. — *February 21, 2012*

☽ Whitney Houston, *1963 - 2012*

This has been a week of tributes to Whitney, so I will add my voice. She was born on August 9, 1963, at 8:55 pm, in Newark, NJ. As I sometimes get stung with unverified birth times, I tried to trace down where Whitney's came from. The best I could turn up in a quick search was Eric Francis attributing it to Lois Rodden herself. Who claimed the birth certificate. Rodden had contacts. They got her data no one else could.

The two questions I have are, Will astrology tell us the cause of Whitney Houston's death?, and, Will astrology tell us if this was her proper time of death? Astrology claims it can do both. For those of you who say these are tacky questions, well, I suppose they are. To me, this is an academic exercise. Let's see if I am skilled.

Whitney Houston has Pisces rising. The ruler, Jupiter, is retrograde in Aries, just outside the 2nd house. This is ordinarily read as someone who eagerly goes after the money, and with the Moon very near by would want it very much. Houston was in fact very rich, but as Jupiter is retrograde, she presumably did not get much satisfaction from having it.

The harder Houston worked, the more money her manager took from her, as Mars rules Aries from Libra, where it is debilitated, but, note carefully, not retrograde. Mars is late in the 7th but shares the cusp of the 8th house. So, according to my usual rule, that a planet outside a house wants to get into it, Whitney's partner (and a manager is a partner) will want to "get into" Whitney's 8th house, where, due to Mars' rulership of Whitney's 2nd, will enable him to get his hands on her money. And she had a lot of money.

Moon conjunct Jupiter, Jupiter being the chart ruler, made her larger than life. Jupiter retrograde, the Moon applying, made her eager to take up the role of a diva. Both of them in Aries made her rash. Ascendant in Pisces meant that it didn't make any difference anyway, since, with Pisces, anything went so why not let the Moon have its way? If it feels good, do it!

Oppositions to the Sun and Moon tell us who the enemy is. When the Moon is beefed up by a conjunction with Jupiter, Mars, the enemy opposite, becomes their target. When that opposing Mars is in Libra, where it is debilitated, there is a sense that "they should have known better" than to annoy the diva. Hence her temper. Sakoian and Acker (*Astrologer's Handbook*) note a tendency for Moon opposite Mars to act out and get drunk, spend money foolishly, etc.

Funny how I couldn't find any of this on Houston's Wiki page. Just as I couldn't find anything but blank praise for Madonna and Lady Gaga.

Whitney Houston was born a Leo Sun. We find it in the 6th house of service, where she wants to "shine." Conjunct Venus (also in Leo) we have someone who is cheerful and upbeat, who likes children. Sun-Venus conjunct in Leo can also be a narcissist. Both planets in the 6th of diet and health, there is the self-confident ease of "I know what's best for me" so far as medicine and food and health are concerned. Leo in the 6th has a weakness for rich, expensive foods, which are rarely good for you. The sixth house is also the house of drugs. Houston once said she did not take crack cocaine, as that was a poor person's drug. Not hers. She was not poor. If someone like this takes up with drugs, it can be very hard to stop, since they will rationalize themselves as being right (the king can do no wrong) and will not listen to outsiders.

Such as Saturn in Aquarius in the 12th. It is opposite Houston's Sun and Venus. It is in the house of institutions, such as, for example, drug rehabilitation clinics. Saturn, representing duty and responsibility, as it is in dignity and opposite the Sun, Houston knew what she should do (so far as her reported drug problem went) but believed herself to be the exception to the rule.

Indeed, Saturn's nagging might have been what led Houston to drugs. There are exactly two squares in Houston's chart, from Sun and Saturn, both of them to Neptune. Saturn represents father – teacher – elder – authority – government – clinic doctor – friends and associates – (Saturn wears a lot of hats) who pointed to the one thing that Saturn and the Sun both have in common — the square to Neptune-as-drugs — and said, Don't go there! You'll be sorry!

Which Houston refused to heed. She did not refuse only because she thought herself superior. She also refused because she could not submit. The Sun rules Leo, Saturn rules Aquarius. Houston's Sun - Saturn opposition was an opposition of peers. It was a stalemate. Maybe she took drugs (Neptune) merely to act out, or perhaps she took drugs expressly to thumb her nose at authority. An authority which she well-knew could never hold her to account, since she was its peer.

Notice how these two different oppositions have played out. On the one hand, the Moon and Jupiter have captured the weak Mars and berate him for their amusement, while Mars, on the sly, steals from both.

By contrast, Sun and Saturn butt heads over Neptune, resulting in a drug-addled person unable to help herself. It would seem that Houston had the worst of both worlds: A drugged out reality that hid an entourage that stole from her.

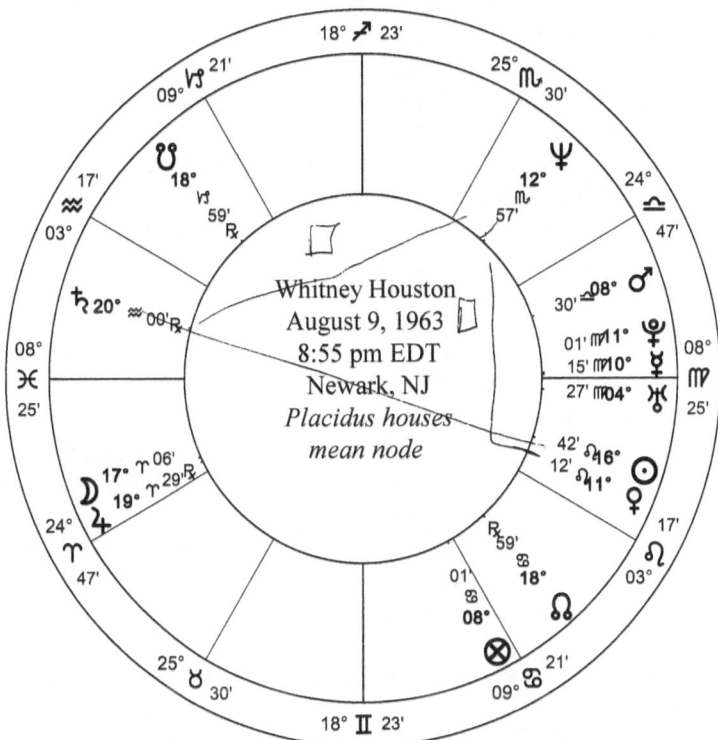

Note also that Houston's Sun and Moon are in trine. Always check the Sun-Moon aspect, as it is the fundamental aspect in the chart. (Even when they're not in aspect, they're still in aspect, just a more frustrating and dysfunctional one.) When the luminaries are in trine, you trust your gut. You go with the flow. Sun and Moon in trine reinforce each other. In square, they irritate and goad. In opposition, they fight. In semi-sextile or inconjunct (which in Ptolemaic terms are not aspects) the head and heart are confused, which, like other aspects, is still an outcome.

Sun and Moon in trine can be a good thing if they help you better live your life, but too often the two luminaries in trine can successfully shut out the world, to their ultimate loss. We are talking here of a woman who had a severe drug problem and who ended up unexpectedly dead at the age of 48. That's not good. That's Sun and Moon as dysfunctional enablers. Watch out for that.

There is a third factor of interest: Mercury/Pluto conjunct in Virgo in the seventh. This is an powerful (Pluto), manipulative (Virgo) partner, one who knows (Mercury) too much.

Regrettably I can find no details of Houston's personal life. It seems to be a secret. How was she able to keep her life a secret? We have only her chart as a guide. So we start over and look at it afresh:

Pisces rising, she could be anyone or anything. The ascendant tells us nothing. Ruler Jupiter in Aries says, fun, outgoing, enthusiastic, except, retrograde, she's not quite.

Sun in Leo is proud, but in the 6th house, he is hidden.

Sun and Moon in trine, you fool yourself as to how the world sees you.

Saturn in the 12th, your problems are well-hidden.

Mutable signs on the angles, you're so busy doing things that nobody notices who you really are. *"Oh Whitney, sing us another song!"* As with Madonna, as with Lady Gaga, Jupiter and Mercury rule all four angles of Houston's chart, and as with Madonna and Lady G, we seem to know very little about any of these women.

Of the two ruling planets, in Houston's chart, like Madonna's, Mercury is stronger planet, being in its own sign as well as being angular. Houston's Mercury, conjunct Pluto, is a manipulative and controlling personal manager (since I can't seem to make Bobby Brown, her ex-husband, into a villain). As with Madonna and Lady G, Houston's Wiki page drones on about the millions of records she sold. If that's all the epitaph she gets, I would judge her life a failure.

So how good a singer was she? I am biased. My forte is orchestral, not vocal, music. My opinion of singers is not a whit different than Alfred Hitchcock's opinion of actors. To Francois Truffaut, Hitchcock notably said, *"In my opinion, the chief requisite for an actor is the ability to do nothing well"* by which he meant the actor should set aside his ego to serve the film in which he appears. For a recording star specializing in pop songs, this means the song comes first. Which is how Frank Sinatra viewed matters. But Frank was a Sagittarius sun. Not a Leo.

Musical ability. From Carter:

Strong Venus and Neptune. Venus in Leo, conjunct the Sun, check. Neptune in square to both. Check. Carter says they should be angular, which they are not, but Venus is co-ruler of the ascendant, which enabled Houston to project beautifully.

Gemini. We find this sign on the 4th house cusp and its ruler, Mercury, well-placed in Virgo in square to it. This enables one to communicate.

Carter says Libra is appreciative rather than executive. In Houston's chart, Libra is the 8th house, which is to say, her partners thought she was very good. Well, we all did. Carter says Saturn-Moon contacts are frequent. Houston had them in sextile, which is good.

16° Taurus/Scorpio: Houston had Neptune at 13° Scorpio. Close.

15° cardinal: Moon at 17° Aries. Nodes at 19° Cancer/Capricorn.

In his summary remarks, Carter stresses Moon or Venus well-placed, as regards to Venus or Neptune, in fire or water. Which Houston had, one way or another. But no one has a perfect chart.

As a singing style, Pisces rising gave empathy, the ability to sell a song. Chart ruler Jupiter conjunct the Moon gave her power. As the Moon in a chart is literally one's own body, a Moon-Jupiter conjunction is literally power, or sheer size, depending on its placement. As her voice lacked nuance, as her arrangements were pedestrian, I confess I never found her of interest. (Mere volume is a liability.) Michael Jackson had Quincy Jones, who gave him fabulous arrangements. Mike had him because Mike wanted that. Just like the Beatles had Phil Spector. Having now delineated three pop divas of middling talent, I think I might have a crack at Michael. He was huge.

We come finally to Whitney Houston's sad death. The 8th house should tell us when. We find 24 Libra on the cusp, we note that Saturn last passed over 24 Libra on Thanksgiving and that it will pass over it twice more in the coming months. This is not important. We are not so poorly made that a mere Saturn transit can end our mortal lives.

I have recently become a fan of Solar Arc directions. There is an option to base them on the Sun's actual travel on the date of birth, rather than the Naibod mean, which is very close. I am in thrall to my theory that directions, in general, represent the rate of decay of the birth vibration.

In the Solar Arc chart set for her day of death, I find the descendant to be 25°25' of Libra. Which is very, very close to 24°47' of Libra that is actually on her 8th house cusp. If that were the basis of a rectification, it tweaks her birth time from 8:55 to 8:58 pm, a mere three minutes.

It is also in line with ancient theory, that when "something" set (i.e., reached the 7th house cusp), you die. But as the 8th house is neither the hyleg nor the alcochoden, its setting cannot be termed fatal. Besides, if it was, life expectancy would never be more than 30 or 50 years, as everyone's 8th house cusp has set by then.

But note the SA 7th house is bracketed by Mercury at 27°16' and Uranus at 21°27', both in Libra. Their midpoint is at 24°21' of Libra. To change Houston's 8th house cusp to match that exactly will move her stated birth time backwards by less than one minute. That's tight.

Are these two planets malefic in Houston's chart? Mercury is the ruler of her 7th house cusp. It also rules her mouth, which is presumably how she came to ingest whatever it was that killed her.

In addition to the 8th house telling us when, the 4th house tells us how, as the 4th rules the ends of things. Note that the ruler of the 4th happens to also be the ruler of the 7th, the partner, and that ruler, Mercury, is physically located in the 7th and in the sign it rules, Virgo. Mars and Mercury hint that Houston's death might not have been accidental.

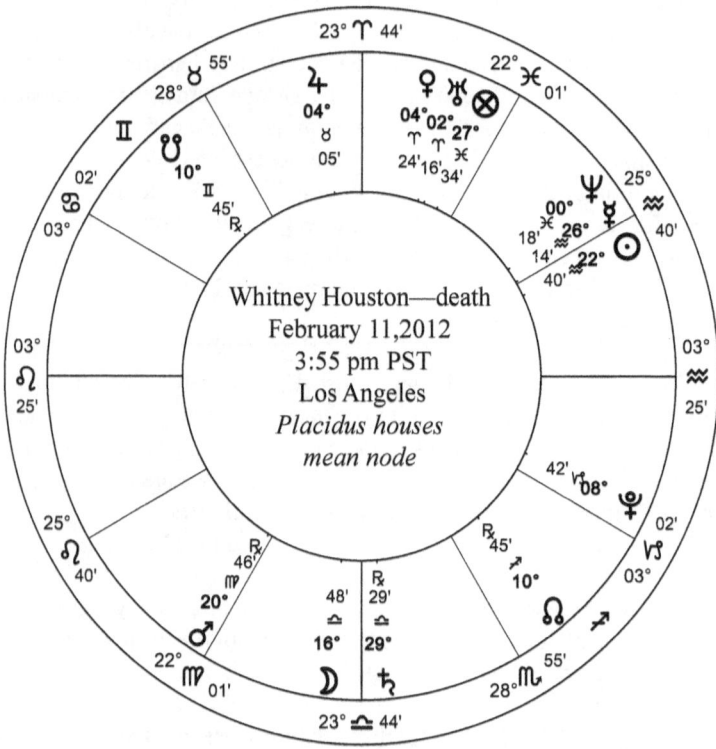

So the question is not *Who prescribed the medications*? but instead, Who gave her the fatal dose, and why? Will that question be answered?

Probably not. Los Angeles is a *very* Neptunian city. Lots of murders are never solved, or are solved at someone else's expense. Remember their primary industry has long been the projection of shadows on a screen.

I take this straight from Sakoian and Acker, who are bold when so many others are timid, or merely cute:

Mars in the 8th house may indicate the likelihood of sudden death; if Mars is afflicted, violent death is possible.

Now that you have this firmly in your mind (Whitney had an 8th house Mars and ended up suddenly dead), I draw your attention — again — to the man who sits in the White House, who allegedly also has an 8th house Mars and who is, in reality, the least likely man in the world to die suddenly. Much less violently. If this is still not clear, I draw your attention to Jack Kennedy, who also had an 8th house Mars. Eighth house Mars people do not live to ripe old ages.

While we can be and often are fooled, Astrology is never fooled. If the chart does not match the individual, then you must rectify the chart until it does, but if that is not possible, then the man in front of you is a fake. Which sometimes happens. In this matter you must never grovel and make apologies. If you have done your work properly and know it to be correct, you may state your results boldly, without fear or doubt.

Was Whitney Houston "accidentally" murdered? Moon-Jupiter conjunct, Sun Venus conjunct, she was *not* suicidal.

Our final analysis of Houston's death is from Abu Ali Al-Khayyat, *Judgement of Nativities*. He gives us a formula for calculating the length of life.

We must first find her Hyleg, the *giver of life*. We first look at the Sun. In Houston's chart, the Sun is disqualified, as it is below the horizon.

Our next candidate for Hyleg is the Moon. In Aries, in the second house, it is in trine to the Sun, which is the co-ruler of Aries. It is hyleg.

We then find the Alcochoden, the *giver of years.* It is the planet which is in aspect to the Hyleg (Moon) and is also the ruler of the Moon's sign, term or face.

For a Moon in Aries, we have two candidates: Mars rules the sign, and the Sun both co-rules the sign, as well as rules the second decanate. In this case we are to take the planet with the most dignities and the closest aspect. Which is the Sun in both cases.

Depending on the Sun's condition, it can bestow one of three possible years:

Greatest: 120
Medium: 69½
Least: 19

As Alcochoden, the Sun bestows its greatest years when it is above the horizon, is angular and is in Leo. It bestows medium years when it is in succeedent houses, and least when it is in cadent. As the Sun in Houston's chart is in Leo and is cadent, and as she lived far longer than 19, I would plunk for 69½.

To this we *add* years if the Alcochoden is aspected by the benefics Venus and Jupiter. As it happens, the Sun is conjunct Venus and trine Jupiter. So we add the lesser years of Venus (8) and Jupiter (12) to get 89½.

From this we *subtract* the lesser years of the malefics if they are conjunct, square or opposed. Saturn is in opposition, so we subtract its least years, 30. Which gives Houston a life of 59½. As this was not her actual life, I speculate that Saturn in its dignity (Aquarius) might penalize with its medium years, which are 43½. Subtract that from 89½ and we get 46. Which is 2009. Now we are close. Whitney lived two years further.

Abu Ali Al-Khayyat says the exact time of death will be when the Hyleg (the Moon) moves, in degrees of right ascension, until it comes to the conjunction or aspect of a malefic. Right ascension is used in primary directions, and I regret I do not have a program that will give me primaries. (I've refused to learn the process as I want something perfectly cut and dried, which primaries have never been.) Converse directions are a rough approximation but in this case do not produce good results. On the day she died, Houston's secondary progressed Moon was at 21 Capricorn, exactly semi-sextile to Saturn. Which isn't the right kind of aspect, but it's an indication.

There are all manner of rules for finding death in a chart. My belief is that we are individuals and can, if we wish, come to our final demise in unique and individual ways. Just as we clearly do with all our other earthly affairs. It is for the astrologer to figure them out, *ex post facto.* — *February 21, 2012*

☉ Michael Jackson

MICHAEL JACKSON was born on August 29, 1958. Alleged birth times vary. I've seen 10:00 am, 7:30 pm, 11:53 pm and 11:55 pm, this last from a KP astrologer running subs. When you can make the subs work, you're doing well, as Krishnamurti is a very demanding system. When I had a quickie look last week I had only seen the 10:00 am chart, which gave him cardinals on all four angles.

My preliminary choice is for the 11:53 pm chart, and for a variety of reasons.

One, whatever he looked like at birth, he essentially made himself into a Gemini pixie. And that's a Gemini ascendant writ large. Did his extreme plastic surgery sometimes make him look grotesque? Sure. But more extreme plastic surgery would correct it. If he lived to 80, he would have gone on getting exotic facelifts. He would have always looked good, at least as long as his health was strong enough to permit the operations.

Second, 11:53 pm puts Leo on his 4th and Virgo on his 5th. Virgo on 5, yes, that's an artist who will micro-manage details, but it's also a liking for young (male) children. Leo on the 4th house cusp, that's living in a castle, no, more than a castle. In one's own country: Neverland. It was 4.7 square miles. Monaco is smaller.

Gemini rising, according to Robson (and me) is an aspect of cruelty. Cruelty to what, exactly? Ruler Mercury debilitated in Leo, Leo representing royalty, royalty being unique, cruelty to himself. Mercury debilitated and wanting to be in the opposite house (the 10th), wanting to look his best, wanting to be regal, kingly, whenever he was in public. Mercury retrograde, he would do anything, and we could see the cracks.

Why did he not have a partner, why was a real marriage denied him? Look at the 7th house cusp. Sagittarius there, he wants some one exotic, someone different, someone fresh. Jupiter ruling from Libra and the 5th house, he wants someone to have fun with, but, Libra intercepted (if the time is right) means he will have problems finding that person, since Libra has no "anchor" in the 5th.

Instead, for partners Michael was pushed, again and again, up against the proper ruler of the 5th, Mercury, debilitated in Leo. Consider that he might well have undergone all the painful surgeries and facial alterations merely to be thought worthy of receiving love, that Mercury, ruling the 5th, wanted to make a perfect impression on his would-be lover. Note also that when we find Virgo on the 5th, Mercury in distress, with Capricorn on the 7th and Saturn prominent, sexual expression can become so onerous and burdensome that one gives up hopes of a real life, and thereafter has sporadic bursts. Saturn on the 7th, he doesn't actually like people.

Saturn retrograde on the 7th, few partners actually liked him. And, regrettably, this is known to be true. Remember the reputation Jackson had? Whacko Jacko. Yes it was unfair, but that it stuck indicates that he was vulnerable. Would you go on a date with him? He's sexy, yes, but he's also, well, not someone you can trust, oddly enough. You expect him to do weird things. Was he a Peter Pan, or was he only pretending? What was real?

MICHAEL JACKSON had the same Sun and Moon, by sign, as our old friend Johann Goethe. Sun in perfectionist Virgo, Moon in hypersensitive Pisces. Goethe was a consummate artist and in fact spent many years reworking Newton's rather crude ideas about color. In Jackson's case, he was just as intense about getting his recordings exactly right.

Imagine a Sun-Moon opposition. Intellectualize it. The Sun in Virgo has detail and fussiness, which he offers to the Moon. But to the Moon, it's exactly what the she does not want to hear. So she rejects it, out of hand. Full Moons say "NO!" a lot, and that's why.

Turn it around, the Moon feels and sees everything. In Pisces it has no filter, it has no defenses. Good, bad, happy, sad, up, down, it's all the same. See the Sun, standing in the other corner. Whatcha got?, the Sun asks. This! and This! and That! The Sun looks and thinks what the Moon has shown him is all perfectly dreadful.

But here's the catch: *The Sun is not an emotional planet.* He has no emotional stake. He does not like or dislike. That's not his function. The Sun shines on everyone and everything. The Sun is regal, it is the ruler of all the planetary citizens, even the ones he doesn't really like. In a literal sense, there can be no such thing as an opposition to the Sun, since the Sun is the very center of all things, by definition. Always has been, always will be.

So when the Sun finds himself in a full moon situation, he must take what the Moon gives him, do something with it, and then give it back to the Moon. But to his surprise, he finds the Moon to be very, very fussy. She must be dealt with on lunar terms, not solar ones. This forces the Sun to invent, and then to reinvent and finally, at wits end, to simply give up and throw the Moon the first thing that comes to mind. And to his amazement, the more the Sun surrenders, the more he simply bats the ball straight back (like ping-pong) the more the Moon is delighted with the result.

In more down-to-earth terms, the full moon person has lots of ideas (the full moon is the ultimate in sheer contrast), but he throws out most of them because they don't "feel right." In Jackson's specific case, his Virgo ideas just did not have the right details. There was always an objection. So he goes back and forth. Will this work? Will that work? Will anything work?

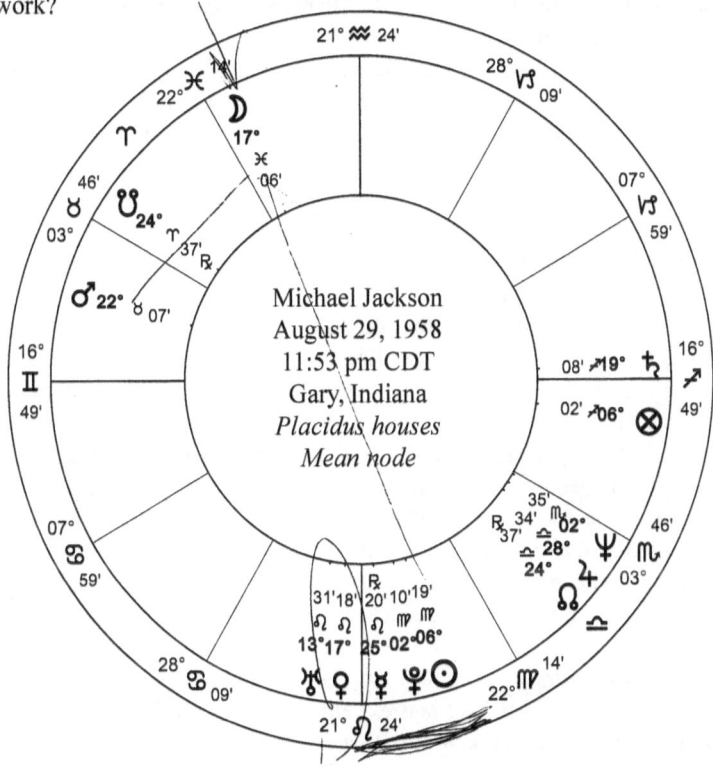

Slowly he grinds out the right idea. Full moons need time to find the one plan where all the details will work. Which will balance perfectly. One that will, more or less, satisfy the Moon. The result will be so far beyond what was expected as to astound the world. The full moon, if it is in any way unleased, will rise to profound, dynamic levels.

But the full moon's sensitivity has a steep downside: They will make you into a recluse. Afraid to leave the house. Every portrait I've seen of Goethe he had a terrified look on his face. The mega-star Michael

Jackson was exactly the same. When he was already quite famous in 1979, he gave a TV interview. His next TV interview was in 1993, 14 years later. Yes, that's Saturn at work (Virgo in '79, Pisces in '93), you can work out the Sun-Moon details for yourself.

We can also work out the relative strength of Sun and Moon. Jackson's Sun is conjunct Pluto, which gives it great power and strength. It is ruled by the chart ruler, Mercury, with which it is in mutual reception. The Sun also rules Venus, Uranus and the 4th house. It is widely trine Mars, and widely square to Saturn, in addition to opposing the Moon. The Sun and Mercury tie the 4th and 5th houses together, with Pluto they are a powerful weight at the bottom of his chart.

Michael's Moon, on the other hand, is tightly square Saturn, which makes him emotionally insecure and defensive, sextile Mars, which is good, and widely opposed to Pluto, which is bad. In addition to the Moon being full. The Moon's two rulers, Jupiter in Libra and Venus in Leo, are both inconjunct, which makes for a *Finger of Fate*. Which, as I think inconjuncts are aspects of invisibility, means his Moon got little practical support from either benefic, even though they are the Moon's rulers. The Moon in the 11th appears lonely, isolated, and vulnerable.

Michael wore a white glove on his right hand, none on his left. Why the right hand? The hands are ruled by Virgo. (The feet are ruled by Pisces.) The Sun rules the right side of the body. If you say that, therefore, Sun in Virgo will wear a glove on the right hand and make that a rule, you will be wrong. Michael developed the glove as a fetish. He kept it on because it helped him express himself in a fundamental fashion. Which, I think, is the secret to fetishes, that they enable the person to express some part of himself, and with this in mind, fetishes may be found to have astrological significance. The glove was white because Virgo is picky and white is clean, it was white because of the Sun's radiance, it was white in reaction to the darkness of Pluto, lurking nearby. It was studded with rhinestones. Sparkly things they are, they shone like the Sun itself.

By contrast, Michael had extreme body sensitivity, which is the Moon. There was a broken nose that never properly healed. His scalp was seriously burned. When, in 1993, he was accused of sexual molestation the victim alleged Jackson to be circumcised. Which, from what I remember of high school locker rooms, would be a good guess. Most American males are circumcised. As part of the inquiry, Jackson was forced to submit to an examination (Wiki says strip searched) and was determined to be uncircumcised. Which was confirmed at his autopsy.

Jackson simply shut down as a result. The case was eventually settled out of court, and as Jackson himself later repudiated the terms, it would appear his insurers did it over his objections, simply to get rid of it.

Michael Jackson

There was much the same result when Jackson's ranch was raided in 2005. Jackson felt it to be a violation. His response was swift and absolute: He abandoned his home of 17 years and went to Dubai. This is a highly sensitive Moon, in this case empowered by an enraged Sun conjunct Pluto.

Daddy. Michael's father, Joseph "Joe" Jackson, is all over the chart, as he was in Michael's actual life. The father is shown by the 4th house cusp and its ruler. In Michael's case, Leo is on the 4th house cusp, which means Jackson's father was a supreme authority of some sort. The ruler of Leo is Jackson's Sun in Virgo in the 5th. Which is the second house to the father, which means Joe Jackson saw his son as a source of money. Sun in mutual reception with the ruler of the 5th (2nd) confirmed Michael as Joe's very own fortune cookie. Sun conjunct Pluto, the father was cruel (documented). Sun conjunct and empowered by Pluto, opposite the Moon, Sun representing daddy, Moon being Michael himself, Michael was square in his sights.

MICHAEL JACKSON was in fact discovered by his mother, Katherine. She is represented by Jackson's 10th house, Aquarius. The ruler is Saturn, in Michael's 7th house, but the 10th when we turn the chart to represent his mother. Saturn retrograde in his mother's 10th house is why Katherine never had her own career. True to Saturn, she tried and tried, but Saturn retrograde, she never succeeded. When Michael was about 5 years old, his mother discovered his talent, but, again, her ruler retrograde, Katherine's efforts to promote Michael, to her husband, to Michael's own family, were ignored. By turning Michael's chart, do we have an absolute portrait of his mother, or his father?

No. Neither. We have only that part that is relevant to Michael himself. Michael Jackson's father was cruel to Michael. Not necessarily cruel to any of his other children. Katherine's aborted career was important to Michael, but not necessarily to herself. When children get together as adults and compare memories of their childhoods, they are often surprised at how much they disagree about the very nature of their own parents. Very likely Michael Jackson's own funeral was the occasion for just such an event.

MICHAEL JACKSON was declared dead on June 25, 2009, at 2:26 pm, in Los Angeles. So let's do a proper analysis. The Sun is not hyleg, as Michael was a night birth. Is the Moon in Pisces hyleg? The Moon is not aspected by its ruler (Jupiter), nor the co-ruler (Venus). It is in square to Saturn, but Saturn has no rulership of the 17th degree. The Moon is sextile to Mars, but, again, Mars has no rulership of the Moon's degree. Nor does the Sun, which is in opposition.

Mercury has rulership of the third term (bound) of Pisces, but is not in aspect to the Moon. Which leaves us with Mars as triplicity ruler of Pisces. (Triplicity means ruling the three water signs. Not the decanate.) If Mars is the Alcochoden, or giver of years, it is both cadent as well as debilitated. It gives minimum years, 15. To this we add minimum years of benefics if conjunct, trine or sextile. As we find none, as Michael lived far beyond his teens, we discard the Moon as hyleg and consider the Ascendant at 16 Gemini as hyleg. The ruler is Mercury, which, in sextile, becomes the alcochoden.

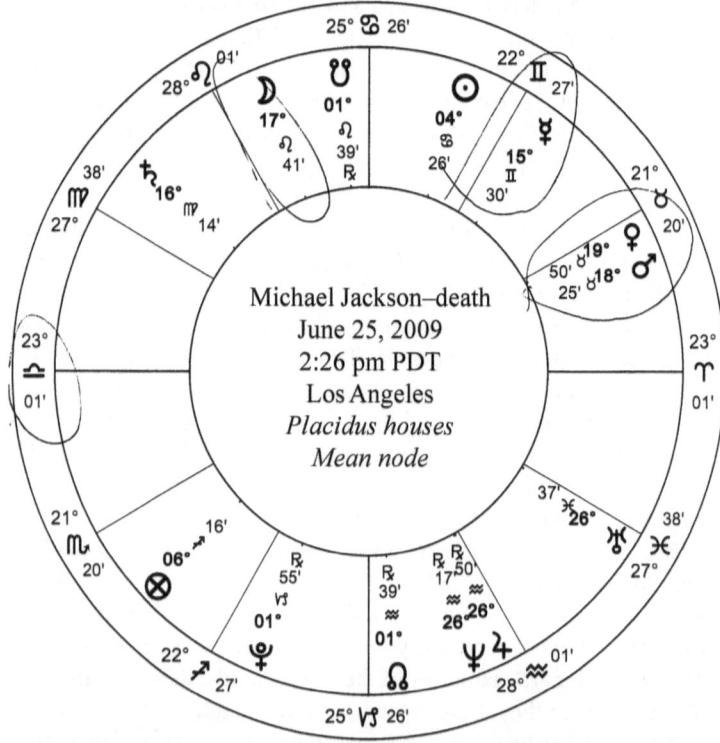

Mercury is angular, which would give maximum years (76), but it's also retrograde, which I think means its medium years, which are 48. Mercury is sextile to Jupiter, a benefic, so we add Jupiter's least years, 12, to get 60. Mercury is not conjunct Venus, as Venus is on the other side of the 4th house cusp, which is an absolute separation. Mercury is square the malefic Mars, so we subtract Mars' least years, 15, to get a final figure of 45. Michael Jackson died at the age of 50. Do you really, really wanna be an astrologer? Are you sure? Compute your probable life expectancy

Michael Jackson

and then decide. I did mine and I've been trying to rationalize it away ever since. I do enough dead celebrities and maybe I'll satisfy myself.

The year 2003, when Michael Jackson was 45, started the final downward slide. He was arrested in 2003, abandoned his ranch in 2005, and then had protracted financial difficulties as well as artistic disappointments. I am not satisfied I understand the traditional formula for life expectancy, nor am I satisfied that excess years are a twilight period, but Jackson's case is suggestive.

In Jackson's death chart, the ascendant is de facto conjunct his natal Jupiter, both in Libra. This is a death that is an adventure, a larger than life (supra-life) voyage with a friend. My private research suggests the death ascendant will become the birth ascendant in the next life, but a rule cannot be made from the one example I have.

The Moon in Leo was exactly conjunct Jackson's Venus, the aspect was in fact exact one hour before death was declared. This is a touching aspect of hope. The Moon, a king, a proud man in Leo, comes to "lift" Venus into itself, to take Michael away. I may be disturbing all of you by describing death as a process, rather than as a door slamming shut, but such is how I see it.

Transiting Mars and Venus were both in Taurus and bearing down on Michael's natal Mars. Mars coming to itself is an old friend (a Mars return, in fact), with Venus, the ruler of Taurus, with it for support. Venus, square to itself, strikes me as trying hard to get it right, to make things nice. Note that *none of these are fatal aspects.* They are mere handmaidens, they are attending angels.

Mere mortals do not die like this. The heavens do not open up to receive them the way they did for Michael Jackson.

Which leaves us with the riddle, the puzzle, the enigma, of the full moon birth. These are men of huge talent, but they are more fragile than snowflakes in April. They are with us for a time, and while they are here they perform prodigious feats, to a mixture of open hostility and mindless adulation. Both of which hurt them horribly. The full moon is by far the most powerful aspect of all, but the universe bestows nothing without a price. The full moon birth pays that price, over and over. — *February 28, 2012*

☉ Daffodils

The daffodils in my flower beds bloomed today! I have not paid attention to them in years past so I cannot say if they were early this year, but it sure seems likely. The tulips and crocuses are pushing up behind them, but will not bloom for several more weeks. I'm not much of a gardener, but it is impressive to see one set of blooms succeeded by a second and then a third.

This past week brought severe storms and shocking loss of life in Kentucky and Indiana, as well as in the surrounding regions. The forecasters were able, a day or two in advance, to alert a general area to the danger of tornadoes, but the specifics, both location and timing, elude them and always will. Astro-weather forecaster Carolyn Egan tells me she is fearful that someday the Weather Channel will take up astrometeorology. I am skeptical. Carolyn, I said, you'll know when they do, because one of their storm models will invariably be right, a week, two weeks in advance. All the official weather forecasters know of AstroMeteorology, have known for many years, but their bias is so powerful that *no loss of life, not even in the thousands, will make them see the light.*

George McCormack said that **tornadoes are products of Neptune.** Here is an extract:

> As the extreme expression of the Venus influence, this planet induces the heaviest downpours in the shortest space of time. During spells of heat, combined with excessive humidity under Neptune's influence, surface air currents ascend vertically, carrying dust particles in large quantities into the upper atmosphere. *Disturbances in the upper air levels are sometimes generated so quickly that a spiral motion is set in operation, forming like an inverted cone. These cloud formations have the effect of vacuums.* As these lowering dark clouds sweep in from the western horizon, preceded by increasing wind velocity, *the intense atmospheric turbulences sweep along narrow paths not more than a mile wide across country.* Such storms are intensified in the lowlands and along waterways, when attended by preceding high temperature. [my emphasis] —*from* **A Textbook of Long Range Weather Forecasting.** — *March 6, 2012*

☉ Rush Limbaugh

RUSH LIMBAUGH, American talk radio icon, has long been a serial slanderer. Parents, don't let your kids grow up to be serial slanderers. Rush has slandered so many over the years, it's hard to imagine that Sandra Fluke would be the one to bring the party to an end, presuming she has.

Sandra Fluke was a 30 year old law student at Washington's Georgetown University. On February 16, 2012, she was invited to speak before a hearing by the House Oversight and Government Reform Committee. She argued in favor of a private mandate for contraception coverage for college students.

On February 29, 2012, Limbaugh said of her and her testimony, *It makes her a slut, right? It makes her a prostitute. She wants to be paid to have sex...* Which ignited a controversy. I have no interest in the details, nor do I want to repeat trash talk. I am interested in the characters themselves.

Rush Limbaugh was born January 12, 1951, at 7:50 am, Cape Girardeau, Missouri. The source is said to be the late Lois Rodden herself, which makes the data AA, presumably from one of her spies in Jefferson City, Missouri, who actually saw the birth certificate.

Rush has 0 degrees Aquarius rising, and because Aquarius makes his Venus in the first house so wonderfully tacky, I'm going to go with Aquarius, not Capricorn, as his ascendant. (Capricorn was only four minutes earlier.) Ruler Saturn is retrograde in Libra in the 8th. My sense of Saturn in 8 is greedy. In Libra there are relationship factors as well, so I look at the cusp of the 8th: Virgo. I now look at the 5th: Gemini, as I would expect. Bounce over to Mercury, which I find in Capricorn, retrograde. It is supposedly in the 11th, but shares the same sign as the cusp of the 12th, which is only four degrees away. In the 12th it will be hidden, and, per Robson (Student's Text-Book), will not hear clearly. (Limbaugh has in fact gone deaf.) Retrograde, it's fighting against that. Sort of like finding yourself on the down escalator when you wanted to go up instead.

102 The Triple Witching Hour

Venus in Aquarius in the first completes the initial picture. A retrograde Mercury ruling 5th is traditionally a liking for sex with boys. Mercury ruling the 8th reinforces. Venus in bizarro Aquarius, Aquarius on the ascendant, in mutual reception to Saturn retrograde in Libra in the 8th, Rush would appear to be a closet homosexual.

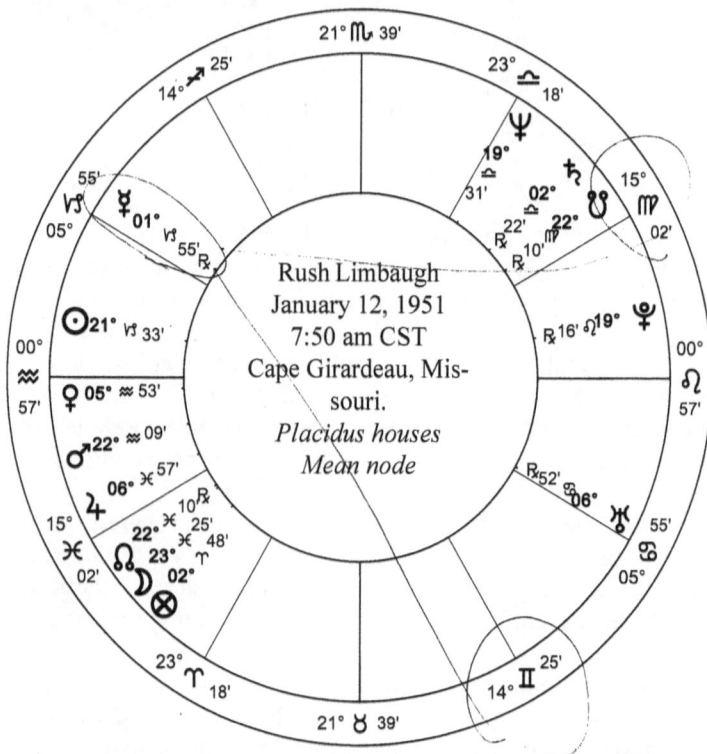

A mere mutual reception, Venus to Saturn, does not indicate gay. Capricorn to Libra, for example, makes one studied and serious and formal and stiff — and luckless, alas! —but not, in itself, gay. Taurus to Capricorn is dull and dense and earthy, but not, in itself, gay. Taurus to Aquarius, that's leaning gay. It's the Aquarian factor that makes Venus-Saturn mutual receptions gay. Aquarius wants to be different. Aquarius acts out. When its ruler is in one of Venus's signs, being gay is an easy way to act out. Anything else would take real imagination. Most people don't have a lot of that.

I know we're supposed to tiptoe and only speak of gays in hushed, reverent whispers, but "gay" is just another chart attribute. Just as you can be born gay, you can be born a cavalry officer or a portrait painter.

So I went to Google and entered, "Rush Limbaugh gay." Up popped a dozen websites that claimed exactly that. So the chart has led me to the same conclusion that others have gotten through hearsay and observation. Remember, parents, don't let your child grow up to be a slanderer. The hearsay about such people!

Since I'm on the "gay thing," is Aquarius a gay sign? No. There is no such thing.

However, the same is not true for the twelve Great Ages, the 2000 year epochs. This is because the Ages, unlike the signs in your horoscope, run full throttle. Every age comes complete with its ruling planet, its exalted planet (if any), and its planets in fall and debility. All at once.

So, for example, the recently completed Piscean Age (ended in 1913) came with Jupiter, the ruler, Venus, the exalted planet, and Mercury, the planet in fall. So of course the Piscean age as a whole was chaotic, as Pisces is naturally unstable. It was religious, because Jupiter is religious. It was artistic, because Venus loves beauty. And it was stupid, because Mercury, the planet of intelligence, was addled. So dumb that nearly 1600 years after its official start, a notably clever man dropped balls off a big tower in Pisa to see which one would land first. Not a sign of intelligence!

The Aquarian age, by contrast, is ruled by Saturn. Mercury is exalted. The Sun is debilitated. Saturn's rulership of Aquarius is different from its rulership of Capricorn. In Capricorn, Saturn is a strict taskmaster. He punishes you today for the sins you were thinking of committing tomorrow. In Aquarius he lurks in the background and pounces at the end to punish you for all the sins you really did commit.

Lacking his constant pressure we think he's gone away for good and act out. Rather like a 20-something who's not yet experienced his first Saturn return.

So in the Aquarian Age, we act out. Being gay is just another way to do that. Mercury, which has no sexual preference, makes us Unisex and thereby muddles distinctions. Females who are masculine. Men who are kind and sensitive. Women who wear trousers. Men who wear drag. Mercury also makes us clever, more so than intelligent. The Aquarian Age, with its debilitated Sun, makes men, who are ruled by the Sun, weak and effeminate. In Aquarius, males fail to be fully masculine. This is simply a trait of the age, but it is a trait which strands women, who, by contrast, have *not* been emasculated. It's not so much that they can't find a straight man, but that they can't find a good one.

Contrast this with the other great gay age, which, oddly enough, is Aries. In Aries, Mars rules, the Sun is the co-ruler. Venus and Saturn are debilitated, and Saturn, notably, acts out. Aries is a hyper-male age in which women — Venus — are excluded. Males consort with each other, women are ignored. Which you can read in Ptolemy.

Unlike heterosexuality, homosexuality has various signatures. Mercury dominant gays prefer boys to women. (Mercury always has an aspect of youth about it.) Why, exactly, is this? Because when the two Mercurial signs are found on the sex houses, 5 and 8, Mercury has to deal with sex. Whatever planet rules the sex houses has to deal with sex and will do so to the best of its ability. In an intellectual chart (Mercury strong by sign, house and aspect) Mercury will study and research and intellectualize sex. Alfred Kinsey, Masters and Johnson, me, etc.

In lesser charts Mercury will give a focus on youth and children, which, as Mercury is not fussy about sex, can be either for boys or for girls. The choice is determined by the sex of the sign Mercury is in, and by any other factors as come to hand. In Limbaugh's chart, Mercury is in female Capricorn, it is retrograde, which makes it retiring (which is a female trait, by the way), it is in a Saturnine sign which makes its proper work onerous and hard and which encourages it to find shortcuts, and it is leaning towards the 12th, a female house, where it can hide. When you see these factors, you judge the native acts, not as a man, but as a woman, and, Mercury signifying youth, chases boys instead of girls. Charts with strong 8th and 12th houses, or cadent placements (3rd, 6th, 9th, 12th) will tend towards the closet and no amount of outing will change that.

This is particularly true of Mercurial sorts. Gemini is two faced and naturally leads a double life. Virgo can never be clean enough and so will not want to be shamed or soiled. Mercury has an affinity for two of the four cadent houses. It's not that cadent houses want to hide, but that they have a hard time expressing themselves. Easier to be a recluse. If you're a gay recluse, you won't want the attention of a formal, gay marriage.

Acting out is another gay stereotype, one that involves hard aspects between Uranus and Venus, very likely with one or both angular, where the result will be on display. In Limbaugh's chart we note that Venus and Uranus are inconjunct (a weak aspect), but while Venus is angular, Uranus is hiding in the 6th. No acting out. Note that Venus has no rulership of the sex house cusps, the 5th and 8th. Uranus can't make Venus act out sexually if there's no place for Venus to do that. In Rush's chart, it's the Venus/Saturn mutual reception that contributes to his homosexuality, not the Venus/Uranus inconjunct.

The "power gay" has Pluto in aspect to Venus, or Mars. Rush has Mars/Pluto opposed, which is an aspect of absolute power, but not sexual power per se, since Mars, like any other planet, can only work through the sign and house it is in, and the signs and houses it rules. In Rush's chart, that's the third house with Aries on the cusp, which is bold, forceful, powerful speech, and the 10th, Scorpio, of public intensity. Neither are sexual per se.

So there is the gay who chases after youth, the gay who wants to be bizarre, and the gay who wants power. There used to be a psychological term, "latent homosexual" which was commonly misinterpreted to mean, wanting to be gay, but not having the guts. For example, Wiki says this:

Latent homosexuality is an erotic inclination toward members of the same sex that is not consciously experienced or expressed in overt action. This may mean a hidden inclination or potential for interest in homosexual relationships, which is either suppressed or not recognized, and which has not yet been explored or may never be explored in fact.

Aside from being so broad that it includes every man who ever lived *(not consciously experienced or expressed),* this is entirely wrong. A latent homosexual finds himself in a power relationship with other men. He is trying to be a "super-man," a hyper-male. All men must submit. All women as well. As a consequence, people are terrified of him. My boss's boss, in food service at the University of Kansas in the mid 1970's, was latent. His name was Forrest J. My boss told me Forrest berated him intensely for being a wimp, for not being "masculine." There are old ballads of frontier roughnecks who dealt with both sexes and who were, in fact, latent, but I cannot quite remember the names.

In his crudeness, Limbaugh's Mars/Pluto opposition gives him such traits. Sexuality in this context is simply another way of exercising power over others. A latent homosexual becomes "overt" when he forces men to submit to him sexually. Not all latent homosexuals have such an interest.

A great many latents will therefore view gays as weak, effeminate targets of opportunity. As both a liking for boys and the need for power can turn up in the same chart — as with Limbaugh — the result can be someone who consorts with gays in private but who, in public, bashes them — and everyone else. Which makes Limbaugh latent to his peers (men his own age), and gay to the younger men he seeks.

All of which is a giant diversion from the week's topic, The Rush Limbaugh Follies. For those who think I am bashing gays, you're right. I *am* bashing gays. As a complex, secretive and powerful group with a large public persona, not only can gays use a bit of bashing now and then, but they ought to expect it. Nobody's sacrosanct.

There are other secrets. Ever hear of the **male** g-spot? It's that part of a man that a woman just can't reach. A gay reminded me of that recently. *Some of us were born this way,* he said, and then with a confident smirk added, *but if mind-blowing sex is what you want, a lot of you can be converted.*

Such is the Aquarian Age, with its weak, feeble men. As women in this Age are neither weak nor feeble, we may see an open female vs: gay war for the souls of wayward males. "Marriage equality" is but a skirmish. When a similar choice was put to women 30 years ago ("sisterhood"), they shot it down.

Rush Limbaugh, with his Sun in the 12th, Mercury trending to the 12th, mass market Aquarius on the ascendant, with Venus nearby, is ideal for radio. His try for TV flopped. Aquarius is too bizarre for TV, the Sun in 12 too retiring, Mercury in 12 cannot hear the others in the room.

Saturn in 8 makes Limbaugh greedy for other people's money. Jupiter trying to get into his second means he will spend lots of money (Jupiter in 2 is a spending, not a getting, placement), which will make him feel personally successful, as his Moon is in Pisces in the second and appreciates the volume. The more money Limbaugh spends, the more his Moon likes it.

Drug use is normally shown by planets in the 6th. In Limbaugh's 6th we find Uranus in Cancer. It is ruled by the Moon in Pisces in the 2nd. So, once he was rich enough, Limbaugh's tendency was to spend his money on drugs and become an addict.

He is unusually fortunate to have his Moon conjunct his north node. The chart says that as long as the drugs continue, he will be fine, which means the episode in 2003 had a cause.

And it did. In July, 2003, in quick succession, Sun, Mercury, Venus and then Saturn passed over his Uranus, Jupiter conjuncted Pluto and opposed his Mars, and Mars conjuncted his Jupiter.

Sun, Mercury and Venus over Uranus were the sucker-punches, the annual I-have-a-great-idea/lust, before Saturn, following right behind, dealt a massive blow. Limbaugh had a Mars return the previous month, he was energized and ready to go. Jupiter hitting Pluto in 7, there was someone else involved.

As for his current problems, the ephemeris shows only a Mars station on his 8th house cusp. The chart of the woman he slandered, Sandra Fluke, tells a different story.

Sandra Fluke was born on April 17, 1981. Time and place are unknown, but as she has long been associated with New York, we can take it as a starting point. The chart is strongly polarized between Aries, with Sun, Mars, Mercury in Aries, with Venus at 0 Taurus; and Libra, with Moon, Jupiter, Saturn and Pluto in that sign. There is little to buffer the oppositions. Neptune in Sagittarius makes trines and sextiles. Uranus makes semi-sextiles and inconjuncts. A most powerful, explosive chart. I wouldn't want to mess with it.

Placing the oppositions (rectifying the chart) should be a relatively easy thing. As Ms Fluke is not already well-known, and as she did not step effortlessly into the limelight when the opportunity arose, we may presume her Sun-Moon polarity is in cadent houses. Our choices are 3 and 9, or 6 and 12.

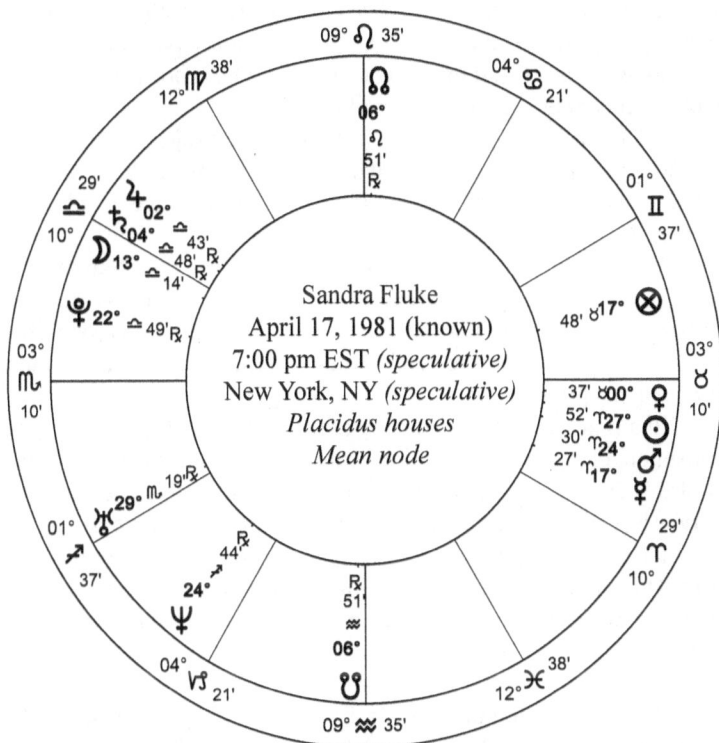

Three and 9 will give a huge intellect and possibly a religious tinge, depending on where the Moon ends up. That's not her.

Six and twelve will be health and slavery (6) and institutions (12), such as hospitals. Throw in Aries-Libra as a contest between the sexes and you get women's health issues and that's her. This gives either Taurus or Scorpio as ascendants. Since she has a neck and has black, helmetlike hair, I'm going to make her Scorpio rising and give her a tentative birth time of 7 pm. When you are rectifying charts in a crude fashion, there is no point in being fussy. Approximations will do, and if you're wrong, you will find out soon enough. Dare to try something!

Moon in 12, along with Saturn and Jupiter, she keeps her feelings to herself. Moon sandwiched between Saturn and Pluto, those feelings are morose, dark and intense. Opposite Sun-Mars, this is someone who has been hurt, and who has hurt others. Like Limbaugh, she has a powerful Mars-Pluto opposition, but unlike Limbaugh, she has substantial backup firepower as Mars is flanked by Mercury and the Sun, and are in the far more powerful sign of Aries. Will Mars oppose Pluto make Sandra gay? No. Women take up with other women for different reasons, but with massive amounts of maleness in her chart, if Sandra Fluke is a homosexual, and if she was born with 6th and 12th house placements, she is even further in the closest than Rush. As Ms Fluke is not a public figure, speculation is rude and an invasion of her privacy.

If Scorpio is rising, Mars is Ms. Fluke's chart ruler. In the 6th, she would be known for her advocacy of health issues. Which she is. She will also be a workaholic. Drug use? With the chart ruler in the 6th, drug use would be on public display, and I did not see that.

Comparing her chart to Limbaugh's, they have their Saturns conjunct, one being 30 years older than the other. Fluke symbolizes Limbaugh's Saturn return, which is an unpleasant time in everyone's life. Which is to say that Fluke herself is a transit to Limbaugh's chart, a transit from an unhappy time in his life.

This is another way of looking at relationships. Is your wife your companion, or is she the human embodiment of the transits on the day of her birth? Transit or relationship? Wouldn't that depend on how long the relationship lasts? It would seem as if Limbaugh and Fluke had a "one night stand," woke up in the morning and decided they did not like each other. And if nobody had shot his mouth off, it would never have become personal. So what's the difference between a relationship and transit? Transits can't shoot back!

Fluke's dual stelliums fall in Limbaugh's money houses. With Fluke's Jupiter-Saturn-Moon-Pluto all in Limbaugh's 8th, the immediate effect of their interaction has been to shut off Limbaugh's source of income. With his Saturn in the 8th, Limbaugh has always been reliant upon others for his money. Money which Limbaugh's Jupiter, not in aspect to Saturn, has always been happy to spend as he saw fit. With the money turned off, Limbaugh will crash through his reserves quite rapidly. If and when he does, it will hit his Moon quite hard, reducing Limbaugh to a little lost boy.

Fluke's Aries stellium falls in Limbaugh's empty 3rd house. The third house is one's daily life. Opposite the 9th house of radio broadcasting, Fluke might actually drive him off the air. It is refreshing to see a chart as powerful as Fluke's.

Parents, don't let your child grow up to be a serial slanderer, for he

Rush Limbaugh 109

is sure to some day slander someone even more powerful than he, and that encounter won't be pretty.

Next week, Ludwig as a study in new moons (I've done so many full moons), or maybe, Putting the Enlightenment In Its Place. Which has surprising ramifications. — *March 13, 2012*

☉ The Heat Wave

In McCormack's *Text-Book of Long Range Weather Forecasting*, I learn that Mars in Leo produces heat. I also learn that Mars produces heat when it is closest to the Earth.

Guess what? Mars was in Leo in October, and then went retrograde in Virgo. Retrograde is when Mars most closely approaches the Earth. The closest approach was when Mars opposed the Sun: March 3. Enjoy the heat while it lasts.

McCormack says to take the chart for the entry of the Sun into the cardinal signs and then look see what turns up in the 3rd and 9th houses, as that's the weather that's going to form due west of you and then drift eastwards.

The Spring Equinox was at 1:15 am EDT on March 20. Moon-Neptune in Pisces fall on the 4th house cusp in eastern Colorado. They are opposed by Mars (still retrograde) in Virgo. East of Colorado, this combination falls in the 3rd and 9th houses, meaning the weather it creates will drift east through the spring.

Of Mars/Neptune, McCormack says, *static electricity is intensified in the atmosphere, particularly during periods of prevailing high temperatures. Temperatures rise, peculiar calms ensue and are followed by squally storms of short duration over narrow areas in the lowlands.* Sounds like tornadoes that intensify as the system moves eastwards and the elevation falls. I haven't yet found where he tells how to time when these storms occur. Let's hope it's not going to be a dreadful spring of tornadoes. They are frightening events. — *March 20, 2012*

☉ The Spring Equinox Chart

An interesting discussion a few days ago on Facebook about the Spring Equinox, which was at 1:15 am EDT, March 20, 2012. Set the chart in Washington.

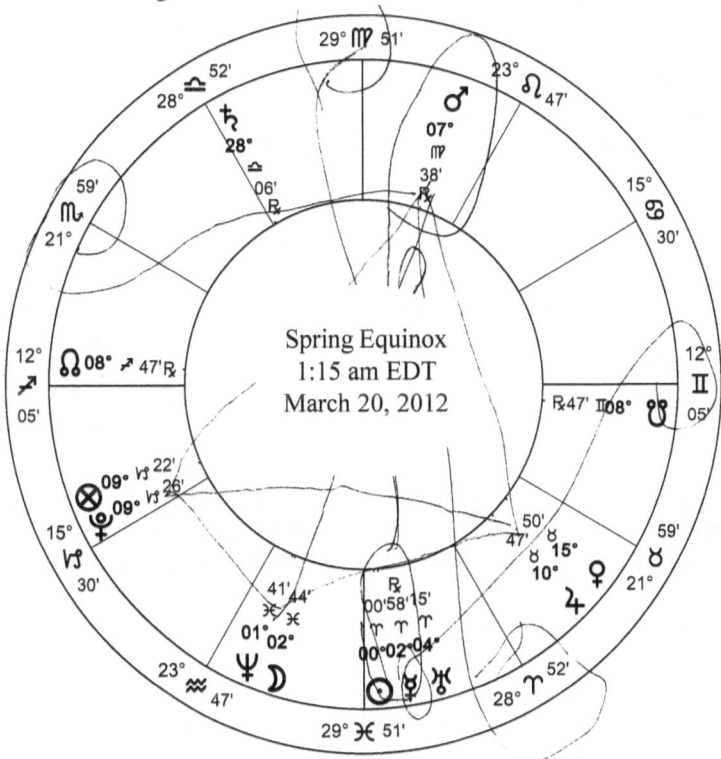

It started with the question that as *Venus is the most dignified planet in the chart, would that mean a successful slave* [i.e. worker] *revolt?*

There's several problems with this already. First, computing relative dignities isn't where you begin with a chart. Computations are useful when the rest of the chart is checkmated, when all your other clues have resulted in dead heat. It's not just that we haven't examined the chart and so don't know if that's true, so much as a deadlocked chart is a fairly rare outcome and when it does appear, it's most often the result of incomplete or erroneous analysis.

Next, the slave revolt. Venus is in dignity in Taurus and does rule the 6th house, which has Taurus on the cusp and the 6th house is slaves, so all of that checks. The problem is that Venus in Taurus is happy with its condition. If there are slaves in the 6th, they're *happy* slaves. Jupiter widely conjunct Venus, also in Taurus, there's a lot of them. How about aspects? Venus is part of a Grand Trine, with Pluto and Mars. Both of these are aspects of power, and in earth signs, can move mountains, but grand trines are notoriously lazy and rarely amount to very much. Moon-Neptune are widely sextile to Venus. Nothing unhappy there.

The Equinox chart has Sagittarius on the ascendant, the ruler is Jupiter, Jupiter is in Taurus where it is conjunct Venus, so in this chart we do, in fact, get very quickly to Venus in Taurus and the 6th, but they're not slaves. In mundane astrology, the 6th is also the military and the police. This chart shows them to be fat (Jupiter inflates) and happy (Venus) and not interested in doing much: Taurus is lazy if not prodded, and there's nothing here to prod. Are the police and military "slaves"? Doesn't make much difference. In 2012, they're all fat and happy.

Then it was observed that Mars was ten signs from the ascendant, Sagittarius to Virgo, and it is. But in D.C., that doesn't make Mars angular and you can't simply smear Mars over the MC and pretend. Angles have meaning. Mars is cadent, in a mutable sign, and retrograde. The retrograde makes him dysfunctional, Virgo makes him fussy and hesitant. The 9th house has him chasing after spooks, one way or another. He can work through the houses with Aries and Scorpio on the cusps, which are the 5th and 12th. Since these houses are not angular and as Mars is not angular and as Virgo is not cardinal, the best Mars can do is be subordinate to the angular rulers, which are Jupiter and Mercury. Mars is square to the nodes, which is not significant as there is no eclipse in sight. Mars is widely square to the Ascendant/Descendant, which, given Mars' relatively weak state, is not significant.

With mutable signs on the angles, with the chart ruler obscurely placed and self-satisfied, with the other angular ruler retrograde, this is a feckless chart, with one exception:

Set the chart in New York and the MC/IC axis becomes 3° Libra/Aries. It shifts the chart rulers from Jupiter and Mercury, to Jupiter Venus/Mars, with Jupiter and Venus conjunct and Mars in trine to both. New

Spring Equinox, 2012

York will have a very different year. A more organized one.

In sum, Sun and Mercury are conjunct in Aries, which is brash, hard speech. Mercury retrograde, talk is backwards, speech circular. Sun in the 4th favors landed interests and the Republicans to some extent.

McCormack says when Sun and Mercury are conjunct in the spring equinox, and Mercury is retrograde, the weather will be *"rain and strong winds. Retrograde Mercury, wind and gales."* Note also the Neptune-Mars opposition, from 3rd to 9th. This is a marker for tornadoes. Storms with tornadoes will form about as far west as Amarillo, TX and then move east. Wanna know if you're going to get hit? Set charts for the lunar quarters for your locale and then check what planets are on the 4th and 10th house cusps. A planet on the 4th sends weather to the north. A planet on the 10th sends weather south. If you get planets in either one, that's your weather for the week, good or bad. So far as April 2012 is concerned, you don't want to find Pisces or Virgo on the MC or IC.

For May weather, set the chart for the Sun's ingress into Taurus, etc.

Politically, the Equinox Moon and Neptune are one degree apart in Pisces. This is fogs and dreams and illusions and lies and deceits all mixed together in a general swirl, interrupted every now and then by thunderous religious sermons — Mars opposed from the 9th.

2012 is an election year. This chart says that nothing will be resolved, that the general muddle will continue unchanged. My guess is that protests will fail to attract a following. *— March 27, 2012*

☉ AstroMeteorology and Astro*Carto*Graphy

Elliot Jay Tanzer emails to say he has a new book on Astro*Carto*Graphy which he's been selling at the AstroNumeric website, as they've long been big on astromaps. Includes a one hour DVD. He's been selling it for $50 or $37 or something. When he gets copies to Jack at the AFA, I'll stock it and tell you more about it.

Tanzer's thesis, with which I broadly agree, is that we are attracted/repelled to those places, peoples and cultures which have favorable or unfavorable lines running through them. My problem is that this has never worked for me personally. We prove astrology by first proving it to ourselves, so, for me, ACG failed. My suspicion is that full moon births, such as mine, are an exception. I am also wondering if powerful placements on the MC/IC axis make one hypersensitive to land and so unable to profit by moving here or there. Anybody out there with similar experiences?

Astro*Carto*Graphy and AstroMeteorology ought to be a perfect fit. McCormack's system (see *A Text-Book of Long Range Weather Forecasting*) is largely MC/IC based, and those are the straight vertical lines on an ACG map. It should be a simple thing to say that, for example, the Mars lines will produce hot weather, Venus rain, Saturn cold, Jupiter pleasant, etc. Since McCormack works with 30° and 45° based aspects, an enterprising programmer could color-code the lines when those planets are in aspect to other planets. For example, when Venus is sextile to Mars, those two lines would be in pink to indicate a sextile between them. You would see a pink line, you would know to look 60° (four time zones) to the left or right to find the other pink line. When planets are in aspect to more than one planet, the line could be dashed, alternating red and green, or red green and blue, for example. Weather forms along these lines and moves east, according to topographic features (rivers, mountains, etc.), which can easily be shown as well. Presuming the world ever has money again, this would be a drop-dead fabulous way to plan a vacation, where the weather was guaranteed to be perfect! — *April 3, 2012*

☽ Ludwig.
The Intense Isolation of the New Moon

Caution: This essay says many strange and bizarre things about Wolfgang Amadeus Mozart. His fans are strongly warned.

Ludwig van Beethoven is an example of a New Moon birth. He displays the characteristics quite well.

As everyone knows, Beethoven was born on December 16, 1770, in Bonn. He was the second birth to his parents, but the first to survive. He was given his deceased elder brother's name, Ludwig, which happens to have also been his paternal grandfather's name. The elder Ludwig was a distinguished player in the Bonn court orchestra. Beethoven had two younger brothers, neither of whom amounted to very much.

Beethoven's time of birth was not recorded. In a remark to me a quarter-century ago, the late Edith Wangemann said that quite often birth times were recorded in parish baptismal records, but not in this case. In 1996, Noel Tyl published *Astrology of the Famed* (out of print), a book with proposed rectifications of a number of people, Ludwig among them. Tyl gave Beethoven's ascendant as Pisces, justifying it with Solar Arcs. At the time I disagreed, but had done no work of my own.

I caution that I am a lifelong fan. By the age of 25 I had memorized, entire, the 9 symphonies, 7 concerti, various overtures, 16 string quartets, 32 piano sonatas, the opera and the second mass, as well as most of the violin/piano and cello/piano duos, in addition to numerous more minor works. In the years since I've cleaned up on the trios. I have the scores to the symphonies, piano concerti, string quartets, sonatas and Missa. I've got more on Ludwig than most Trekkies have on Captain James T. Kirk. You are warned. You are entering take-no-prisoners, hard core territory.

Time of birth. There are many opinions exactly when Maria Magdalena Keverich Beethoven, Ludwig's mother, delivered him:

Edward Lyndoe, writing in *American Astrologer*, May 1970, gave Ludwig a 3:40 am time. This gives 10° Scorpio rising, with his Sagittarius cluster in 2, opposed by Mars in the 8th. "My money vs: your money," which isn't Ludwig. While he carped about money, music was his passion. The first rule of any rectification is that it must show the elemental structure of the native. An early morning birth also messes up the baptismal record which is dated the 17th. Babies were baptized as soon after birth as possible. The only reason for a full day's delay would be if the weather had shut Bonn down. As winter weather sometimes will. Any of you weather forecasters want to venture the conditions on that day, December 16, 1770, in Bonn?

Next up, Ralph Kraum, an expert on old horoscopes, writing in *American Astrologer* for December, 1970, gave 1:29 pm. This flips Lyndoe's chart to give 20° Taurus rising. The Sag cluster (Sun-Mercury-Moon, with Jupiter nearby in Capricorn) is now in the 8th, with Mars opposing from the second. This is Your money vs: my money, which still isn't Ludwig, but Moon in the 8th will talk to spooks. Beethoven was a keen observer. In the Ghost Trio, op. 70, in the slow movement, Beethoven describes a haunting. For much of the movement there is a rumble in the piano's left hand, which mysteriously stops towards the end of the movement, resulting in an eerie stillness. If you've ever been in a haunting, you'll recognized that as the moment when the "ghost leaves." The outer movements describe a man's chaotic daytime life, with an overindulgence in "60 bean coffee," the drug of choice at the time.

Kraum's 1:29 pm puts Saturn (in Leo) in the 5th, which describes Beethoven's relationship with children, though does not quite describe his compositional style, which was more fluid than that. Saturn in 5 is not a lack of creativity (I have it there), but it is a very tightly structured one. Which is not Ludwig. Taurus rising, ruling planet Venus in Capricorn in the 10th. This does not produce temper, nor erratic behavior. So I will move on to the next guess:

Robert Jansky gave Beethoven a 4:11:40 am birth. This is the same as Lyndoe's chart, only with 16° Scorpio rising. Same 2nd/8th polarity. Saturn in the 9th, which makes one dogmatic about religion. By the time Beethoven came to write his second Mass (1820-ish), he was conceptualizing Jesus Christ as a personal friend and brother, which is not to say he was fundamentalist, but that his thinking had deepened and matured. (Drop "Jesus" and I will personally agree with him.) Beethoven was otherwise no more concerned about religion than anyone else at the time.

Marion March proposed 1:00 pm. This is the Taurus rising chart, now with 7° there. Beethoven very much looks like a Taurus rising: Bull-necked and stubborn. Pluto is now exactly on the MC, which is presumably because of his transformation of western music. I think this an *ex-post facto* justification. March smears Gemini and Sagittarius over houses

2, 3, 8 and 9, which does not make for a musician at all.

The final entry in the list at AstroDataBank is 11:03 pm, given to Noel Tyl. I am surprised, as this gives a Virgo rising chart, as I remember a Pisces chart. Checking further, I find this to be a typo. On Tyl's own website, he gives 11:03 **AM** as his proposed time. Of his rectification, Tyl says,

> *Ludwig van Beethoven had Venus peregrine ruling his 3rd and the 8th; affairs of these Houses would run rampant in his life-experience [His brother and sexual frustration]. Additionally, an Earth Grand Trine told of his severe isolation. When Beethoven's brother Kaspar Karl died, he left his 9-year old son, Karl, in Beethoven's care. Beethoven's complex celibate neurosis over lineage and family life led to tragic aberrations, legal considerations and further loneliness for many years, even to the point of Karl's two suicide attempts. As well, the Master's Mars opposed his Mercury, Sun, and Moon.*

Which is mostly true (Beethoven in fact stole Karl from his mother, over both of their heated objections), and to Tyl's credit, is rarely cited by Beethoven's many biographers. I've made fun of Tyl in the past, but he is one of the few who are brave enough to put his opinions on paper. Tyl has extensive musical training and although he's not showing it here, what he is showing is his lifelong admiration for the man under study. Beethoven's sad adventures with his nephew (Ludwig destroyed Karl, he was that brutal) are the best kept secrets of his entire life. Tyl did extensive research and my hat is off to him.

But I am going to pass on Tyl's rectification as well. Tyl puts mutable signs on all four of the angles, which is an aspect of chaos. Beethoven's life was not at all chaotic. It was in fact rather fixed. Beethoven did not crave the limelight, which is what one would reasonably expect from Sun and Mercury both in the 10th. Nor did Beethoven travel. His one big adventure was leaving Bonn and coming to Vienna in late 1792. He often talked of travel, most particularly to London at the invitation of Ferdinand Ries, but in fact seems to have never ventured further than a day's coach from Vienna. In his 20's and 30's Beethoven was a man of few possessions, indifferent to his whereabouts and sensitive to criticism, and so often moved from room to room, on whim. Which seems to have misled many.

Aside from Tyl, there is a consensus that Beethoven had a fixed sign rising. I just don't think it was Taurus or Scorpio.

Go to the next page. Look at the illustrations:

On the left, Beethoven Out Walking, a sketch, c. 1820, by Josef Daniel-Bohm, from the Beethoven-Haus, Bonn, as shown on pg. 227 of the *Ludwig Van Beethoven Bicentennial Edition*, a book, published jointly by the Beethoven-Archiv-Bonn and DGG, 1970.

On the right, David Anrias's sketch of the first decanate of Leo rising. Ludwig's face is piggish, but the stance and glare are identical.

Ever a lover of averages, I have set a chart for 7:30 pm, December 16, 1770. An evening birth. Saturn on the ascendant, it was difficult. Mother and child may rest for the night. The priest will be sent for in the morning.

In the proof of the rectification, let us start, where we should always start, with life expectancy. As before, I am using James Holden's translation of *The Judgment of Nativities,* by Abu 'Ali al-Khayyat.

Hyleg. The Sun cannot be hyleg because the birth is at night. With Leo rising, the rising sign, ruled by the Sun, cannot be hyleg. So we look at the Moon, which we find in 21° Sagittarius. To be declared hyleg, it must be aspected by one of the various planets which rules 21° Sag. Jupiter is the overall ruler, but Jupiter is in the next sign, so no aspect. Jupiter is also the nighttime triplicity ruler, still no aspect. Mars rules the term, and is opposed. Saturn rules the face, and is trine. Which makes the Moon hyleg, or giver of *life*.

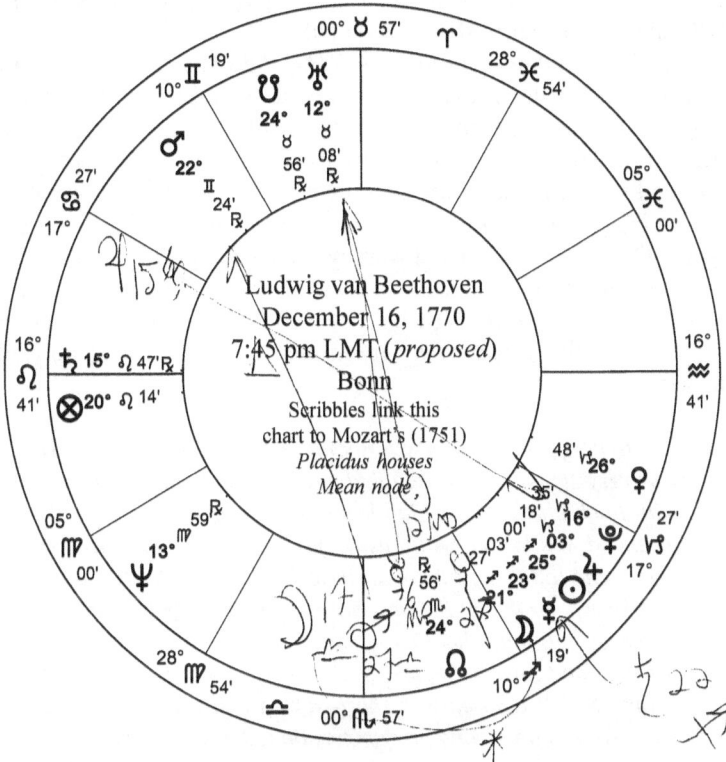

Alcochoden. We next look for the alcochoden, the giver of *years*. The alcochoden must be a ruler which is in aspect to the hyleg. We have already identified Mars and Saturn as being rulers in aspect to the Moon. We go first to Saturn because, in a Leo rising chart, Saturn is angular. Going to the table on pg. 9 in the *Judgments* book, the life for a man with angular Saturn as alcochoden is 57 years.

To this we add the minor periods of Jupiter or Venus, if either is in trine or sextile. They are not. *From* this we subtract the minor period of Mars, if square or opposed. It is not. Which makes 57 our final answer.

Ludwig van Beethoven died on March 26, 1827, in Vienna. He was 56 years, 3 months, 10 days old. It was his 57th year. Sometimes this stuff is just creepy. Of itself, this is not a proof of rectification, since in charts where Saturn is alcochoden, it will produce 57 years if on any of the four angles. It does hint of a night birth, that the Sun is not hyleg.

I am too close to the subject and, like many others before me, will compulsively see Beethoven as I want to see him, which in my case is far from the traditional view. Saturn in Leo in the first, to me, is a bully. Saturn in Leo in the first, ruling the 7th house, no woman will take him as her husband. None did.

Sakoian and Acker (*Astrologer's Handbook*) describe Saturn in Leo as dictatorial and dogmatic. Such natives can be stubborn and rigid. They require attention and respect from others, with a desire for power over them. In the first house, Sakoian and Acker describe Saturn as cold and unfriendly. A difficult childhood is noted, with a general mistrust of others, which produces further alienation. They note two distinct physical types: One of them, short and swarthy, with dark eyes (pg. 193).

A contemporary description of Beethoven, taken from the *Journal of Music and Letters*, vol. VIII, no. 2 of 1927 (the centennial of his death), in an article entitled, *Beethoven's Appearance*, by W. Barclay Squire, reads, in part,

> *In height Beethoven was not more than 5 feet 5 inches. He was very strongly and broadly built, his head unusually large, with thick black hair, which began to turn gray in 1816. The forehead was broad, with bushy eyebrows and rather deep set eyes. The latter are variously described as black, bluish-grey and brown: probably the last epithet is correct, but all the evidence goes to show that they changed in expression in a remarkable way and gave the impression that they were bluish-grey. Until about 1817 Beethoven was short-sighted and used eye-glasses. His nose was broad and his complexion red and swarthy...*

Which not only confirms Sakoian and Acker as to swarthy and black, but, with his thick mane of hair and large head, Leo rising in general. I regret I do not know what the average male height was in 1800, I suspect 5'5" to be more or less average, or only an inch or two short if not. Note that if Ludwig was short and customarily wore lifts, he would appear to be taller as a result. This description is how Beethoven appeared in the street, it was not taken from a medical examination.

When we look at someone we see the rising sign and then whatever planets as may be in the first house. With the Leo rising chart, when we get past Beethoven as a hostile bully (Saturn in 1), we are immediately transported to his chart ruler, the Sun in Sagittarius, in the 5th. We find someone who is creative, as well as striving. Is he musical? Let's go again to Carter's *Encyclopaedia of Psychological Astrology:*

Under *Musical Ability*, Carter starts with strong Venus and Neptune. In Beethoven's chart, unless you put Venus on an angle, it is weak, in Capricorn and in the 6th in my rectification. In fact Beethoven's music was of great power but rarely of great beauty. The supremely beautiful works were few. The two greatest, in my view, being the second Mass (the Missa) and the Sonata op. 109. True to the Venus in Capricorn temper, beauty is a function of form and is only employed where necessary. Never for its own sake.

Neptune is more strongly marked. It is the apex of a T-square that includes Mars in Gemini and Moon-Mercury-Sun in Sagittarius. It is also makes a grand trine to Jupiter and Venus, both in Capricorn, with Pluto sitting very nearly at the midpoint of those two, along with Uranus in Taurus. Of these two groupings, Neptune is the only planet in both.

Carter says fire and water will both be prominent in a musical chart. Beethoven has fire (enthusiasm) to spare, but no water whatever. He is consequently not emotional and may quite frankly be described as heartless. Aside from his psychological destruction of his nephew (which occurred under the full glare of the Viennese police, who in fact arrested him at one point, c.1819 for child abuse), Beethoven's petty cruelties were recorded by Ries (1838), who himself endured a number of them. Further evidence was given by observers at his early salons. Beethoven's playing was said to bring people to tears, but he himself remained unmoved.

Carter says Saturn will be in aspect to an angle, which is true here, as it is on the ascendant. Carter says Saturn gives acute hearing. Before he went deaf, Beethoven had the most acute hearing, or, to be more precise, exceptional skills in observation, which went far beyond mere hearing. Carter says Moon-Saturn contacts are common. Beethoven had them in trine.

Of Carter's degree areas, Beethoven lacks 16° Taurus-Scorpio, though Uranus at 12° Taurus comes close. Of 15° cardinals, Beethoven has Pluto at 16° Capricorn. Beethoven has nothing at 24° of cardinals, though Venus is at 26°. (Carter's degrees are mean, not necessarily precise.) Beethoven has nothing in the ends of Leo-Aquarius, nor anything at the beginnings of Virgo-Pisces.

With a strong 5th house chart, why did Beethoven not have his own children? Or did he?

This is a judgment of fertility. Fertility relates chiefly to water signs. Lacking any water, Beethoven was infertile. The likelihood of marriage can also be judged from water placements. The more water, the more likely, and at an early age. Beethoven was a lifelong bachelor. According to Ries, Beethoven had many girlfriends, but, according to me, a bachelor myself for too many years, "many girlfriends" in practice means many transient relationships. With three planets in the 5th, there was always another woman, one way or another. A fire-earth combination, such as Beethoven's, tends to be brusque and lack finesse. Fire scorches the earth, it brings it no peace.

With three planets in the 5th, while Beethoven was unable to sire children, he would have a strong interest in children and would tend to impose his own grandiose (Sagittarian) ideas on them, and do so nonstop. With Mars opposite from Gemini (a cruel sign, according to Robson and me), childish pranks (opposition) would be brutally put down. This is a very dangerous chart for children, as Karl van Beethoven learned to his horror. Tyl's note that Karl attempted suicide twice while under his uncle's care is, I regret to say, all too true. Karl lived to be 52.

In Beethoven's chart, his brothers are shown by his third house, which, Leo rising, is empty. Virgo is on the cusp is fussiness with brothers, which is true. Most of Beethoven's opus 40-49 works were early juvenile pieces which one brother or the other found and sold to a publisher for their profit, Beethoven himself having abandoned all but one or two of them. This meddling is shown by the ruler of the 3rd, Mercury debilitated in Sagittarius in the 5th of creativity, opposed by Mars in Gemini.

Brother Kaspar's children are shown, in Beethoven's chart, as the 5th house from the 3rd. What is the 5th from the 3rd? It's the 7th, ruled by a debilitated Saturn in Leo in the first. The luckless nephew not only got caught by an uncle with a cruel 5th house, but was additionally seen as a prize possession, an ideal "partner." *Beethoven's nephew Karl was the only person, male or female, who ever lived under the same roof with him.*

I am just now tweaking my rectification. A Leo rising chart can have either Aries/Libra as MC/IC, or Taurus/Scorpio. Both have Mars retrograde in Gemini, and Venus in Capricorn as the rulers, but the shift reverses them. 7:30 pm puts aggressive Aries on the MC, placid Libra on the IC. 7:45 pm puts placid Taurus on the MC, sneaky Mars on the IC. I am probably indulging in my own fantasies, but I like Taurus on the MC. So many people view Beethoven as a Taurus. The sign must be prominent, one way or another. *Wrecktification* by accident (the usual method) produces erratic results. Could it be that rectifiers mistake midheavens for ascendants?

I am of the opinion that a number of Beethoven's previous lifetimes were spent as a monk, locked in his cell in a monastery or ashram, shut away from people, until he had entirely sunk into himself. I am not so certain I have not followed a similar path.

In early 1787 at the age of 16, Beethoven travelled to Vienna. Wiki guesses he arrived in January and left in March or April, upon receiving news of his mother's final illness. Of that time, Mozart was in town for some six weeks. There is no evidence the two men met. Beware of the many people who would like them to have met, and who have invented fanciful stories to that effect. It is very likely that while in Vienna, the young Beethoven, a stranger in town, sat alone in his room, went nowhere, met no one, and when his time had run out, returned home to Bonn empty-handed. Where he stayed for the next five years.

Back in Bonn, Beethoven did not write follow-up letters, did not compose new works, was not visited by the eternally wandering Mozart whom he had either impressed or annoyed (take your pick) and most importantly, did not start a solo tour as a performer. These are all things that young musicians customarily do. Instead, Beethoven simply sat in Bonn.

Beethoven returned to Vienna late in 1792, on the eve of his 22nd birthday. At the time he thought he was 17, from his father's manipulation of his age some years earlier in a failed attempt to pass him off as a prodigy.

Even if his official age was wrong, biology cannot be denied. At the age of 21, Beethoven had no profession, apparently did not work, had only a few juvenile compositions to his credit (the op. 40 stuff, plus the Bonn sonatas, etc.) was not married nor engaged to be married. He was, in a word, unambitious and stranded in Bonn. From many descriptions of those who knew him later in Vienna, he was unsociable and lacked people skills. Yet in 1792 he suddenly moved to Vienna and was the talk of the town and thereafter never lacked for commissions. What happened?

How did this man come to the attention of Viennese nobility and rise to the very top? What was different about 1792? I am of the opinion that Beethoven had help. Lots of it.

I am of the opinion that in the fall of 1792, Mozart visited Bonn on his own business and while there, made the acquaintance of Beethoven, who impressed him so much that he took him with him when he returned to Vienna, leaving the young Ludwig in the care of Baron van Swieten, a friend of Mozart's, who became one of Beethoven's earliest Viennese friends. For this rescue, Beethoven was grateful for the rest of his life.

You will say, *Dave, everyone knows Mozart died the previous December* and I will reply that **everything you know about Mozart is wrong**, starting with his date of birth. Look again at the data attributed to him: January 27, 1756, 8:00 pm, Salzburg. Moon and Pluto are exactly conjunct. Saturn sits ominously behind the Sun and Mercury, which are tightly conjunct. Before spring had arrived, Saturn had conjuncted his

Sun, and then conjuncted Mercury, his chart ruler. Wolfgang Gottlieb Mozart, for that was his name, was most likely dead by April.

In his place, his father substituted his second son, Johann Amadeus, born November 4, 1751, passing him off as a prodigy. Prodigies *always* have fake birth data, they are *always* older than they claim. Mozart's accepted birth data was supplied by his father, Leopold Mozart himself. Leopold was, by profession, the frustrated 4th violinist to Count Leopold Anton von Firmian, the ruling Prince-Archbishop of Salzburg. Leopold was 4th violinist from 1747 until around 1762, being repeatedly passed over for promotion, at which time he took up the promotion of his children (Wolfgang and Nannerl) as prodigies.

The date of Mozart's death, December 5, 1791, in Vienna, is equally false. It is sufficient to remark that no body, and no known cause of death, despite more than 220 years of intense and careful research, leaves no other conclusion than Mozart's death was faked. The oft-repeated story of the public hearse is most revealing. The public hearse did not snatch away the bodies of the poor, for that was not its function. Every house in Vienna, and therefore, every one of its citizens and guests, were assigned to one or another of the parish churches. When there was a death in the family, the local parish was notified. Who sent their own hearse, took the body to the parish church, where rites were performed and the body interred the very next day. This was regardless of ability to pay, and in 1791 the Mozarts were neither poor, nor obscure. In the final year of his accepted life, Mozart gave numerous large scale public performances. He very liked died (if he did die) with money still owed him.

Constanze Mozart, née Weber, Mozart's widow, knew the routine full well, as in the years previous she had *buried four of her own children*.

The public hearse was to dispose of the remains of strangers who had died the previous day and whose bodies had been removed by the police and were being held at various locations in the city. The public hearse had no other function.

So in the story of Mozart's death, why is there a public hearse? I believe it was because Mozart had run afoul of the Viennese authorities, who, panicked there would be a revolution similar to the one underway at the moment in Paris, were rounding up undesirables and removing them from the city. In 1791, with the sister of their Emperor on the throne in France (Marie Antionette was Austrian, not French), and she slandered and under arrest, fear of revolution, either for or against the Emperor, was palpable in Vienna. Mozart's librettist, Lorenzo da Ponte, took the hint and fled. After leaving Vienna, da Ponte had a varied career. He is buried in Brooklyn, New York.

When Mozart refused to get the hint (he, like Marie, presumably thought things would blow over), some bureaucratic functionary came up with the bright idea of commissioning Mozart to write his own funeral music. As soon as the music was delivered, the city would declare Mozart dead and play his own music to prove it, which can only be described as bureaucratic idiocy. When Mozart took the money but produced nothing, the city went to Antonio Salieri for the music. As soon as Salieri was finished, the city made its move. Early in the morning of December 5, 1791, the public hearse stopped at the Mozart apartment and the very living, very alive Wolfgang Amadeus Mozart was physically seized and hauled off. The city left behind an official, signed, death certificate: **Wolfgang Amadeus Mozart died of tuberculosis.** Look it up.

The very living Mozart had corpses for company that night. The next day, Mozart, still very much alive, fled for Prague. Shocked at what he had been an unwilling part of, Salieri destroyed his funeral music and blamed himself for Mozart's tragic ending. In piecing this together, I have read much shoddy, third rate scholarship. There is a great deal more to the story. Some day I will write a book dedicated to it. Some details of Mozart's subsequent life were given, unwittingly, in Agnes Selby's 1999 book, *Constanze, Mozart's Beloved*. To which you are referred. After his reputed death, Mozart in fact can easily be found. He was using the name Georg Nicholas Nissen, he was posing as a Danish diplomat, and was living openly with his wife, Constanze. As a foreign national living under diplomatic protection, Vienna could not touch him. That this was a fake identity is shown by the existence of another Danish diplomat, Nicholas Nissen, who was instrumental in freeing US marines from Tripoli in 1805. Evidence points to Gottfried van Swieten as the source of Mozart's fake identity. More work needs to be done.

In recent years a team of Italian researchers, led by Georgio Taboga, have critically reexamined Mozart's compositions. They claim that many of them were borrowed or stolen from other composers of the era. This reminds one of similar complaints made against Mozart while he was alive, by Johann Christian Bach, in London, among others.

Taboga believes Mozart's many piano concerti were actually composed in Bonn by its Italian kapelmeister, Andrea Luchesi, presumably starting with his arrival in Bonn in 1771, their subsequent attribution to Mozart being a mistake. In support of this, Taboga notes that Luchesi's archives were "disturbed" in 1792, which presumably resulted in Luchesi's compositions being reassigned to Mozart by means of the Elector, Maximilian Franz, but as Mozart was believed to be dead, Taboga could not explain why Maximilian would do this.

With Mozart being alive, living under a new name, and now in search of a new career, it would seem that Maximilian, originally from the court in Vienna, wanted to plunder the archives in Bonn to enrich the musical life in Vienna and used the "deceased" Mozart as his means.

Mozart came to Bonn to copy Luchesi's music (Mozart's scores are unique in that they have few if any blotches in them, very much as if they were copies) and in the process makes the acquaintance of Beethoven, who by that time must have been very restless. The two men become fast friends. When Mozart leaves, he takes Ludwig with him and installs him in Vienna as his protege. Much as Mozart's father had made Mozart a protege many years before. He left Beethoven in the care of Gottfried van Swieten, a diplomat of Dutch origin who resided in Vienna. Between these two sponsors, doors were opened, arrangements made and the very unsociable Beethoven became an overnight success.

In gratitude, Beethoven gave Mozart a number of chamber works, which today are an accepted part of Mozart's œuvre, but which are easily distinguished, as Beethoven had a most unique style. Beethoven dedicated his first symphony to van Swieten, as his first choice, Maximilian Franz, had recently died.

Beethoven's new moon meant he was tightly wrapped into himself. His thinking and his feelings were the same. His antisocial first house Saturn pushed him further into himself, such that he was all he cared for. As a composer, he wrote write what he wanted, the way he wanted. Since no one liked him, he was indifferent to public opinion.

As a child in Bonn, no one challenged him, no one made him do anything, save his father, whom he disliked. Luchesi was his teacher (not Neefe), but when Mozart came to steal Luchesi's works and become Beethoven's patron, Luchesi had to be dropped from the official biography, as it may be presumed a wronged man will not remain silent. In other words, we cannot understand Ludwig until we understand Mozart. The two men are linked.

There is an old story that Beethoven was initially trained on the organ. The organ is a valve instrument, valves are either open or closed, the touch must be firm to avoid the pipes "squeaking," requiring a hard, "staccato" touch. A similar touch can used on harpsichord, spinets, virginals and other plucked keyboard instruments. But not the piano.

Beethoven at first played the piano as he played the organ: He hammered on the keys, until, as the story goes, he was shown the proper touch. At which time he quickly grasped the underlying concept, and his playing changed overnight.

True or not, the story shows a man who must be shown the way, at which time he will adapt easily. This is the isolation of Saturn in the first. The adaptability is his Sun-Moon-Mercury in mutable Sagittarius. This pattern continues throughout Beethoven's life.

Beethoven's early compositions, his "first period" are crude. He has great talent but does not know what to do with it.

The Sonata op. 10 no. 3, and the Pathetique Sonata (op. 13) changed this. The Pathetique was modeled after Mozart's 14th Sonata/Fantasy K.475, which, as Wiki says that Artaria published in 1785, was presumably written by some third hand. (I would otherwise presume that Beethoven wrote K.475 himself, either before or after the Pathetique.)

In these two early works, Beethoven creates a new, simplified style. He uses piano sonatas as experiments as he can control the process from start to finish. At first he goes for simple dynamic expression, but very quickly turns into a story teller. The result was his second, or middle period. In point of fact, all of Beethoven's middle and late period works tell stories. They all have programs.

Why did Beethoven not give his works titles? A simple answer: He did not want performers playing to the title, which has the unfortunate effect of trashing the work as a whole. In fact, Beethoven titled his works only when they could not be played properly without the title. The famous sonata, *Les Adieux,* op. 81a, is a case in point. It cannot be played properly unless you know the meaning of the first note of the third movement. The Sixth Symphony has titles as Beethoven wanted to write discrete, programmatic music and so included a number of passages that needed explicit titles. I have puzzled out a few of his programs, but I digress. Beethoven was not alone in refusing to give programmatic music explicit titles.

Beethoven's career continued until 1812, when Beethoven's patron, Mozart, was arrested by the Danes (at the instigation of the real Nissen, who had retired, returned to Copenhagen, and discovered the fakery) and taken to Copenhagen, where he languished until 1820. The Viennese authorities were powerless to stop them. Having posed as a Dane, Mozart had to accept Danish law. Mozart's trial in Copenhagen should be part of Danish record. In 1820 he petitioned the Danish King, who pardoned him, whereupon Mozart returned to Salzburg, where he died on March 24, 1826. He is buried next to his wife, Constanze, who passed in 1841.

For Beethoven, the loss of his patron was dramatic. After 1812, he virtually ceased to write music. His commissions either stopped, or he no longer filled them. Whatever you may think of Mozart — or my opinion of him — he had a most finely developed ear. Mozart may have stolen much of the music attributed to him, but he only stole the best. I can well imagine Mozart pouring over Beethoven's sketchbooks, encouraging the very specific, very lucid style that we know as Beethoven's second period. Curiously enough, once Mozart had left town, in his "third period"

Beethoven reverted to a mature version of his first period, which peeks out in a tiny handful of compositions, among them the Largo from the sonata in D, op. 10 no. 3, and the Bagatelle no. 7, in A flat, op. 33. This style is highly abstract and very forceful. It is masterful, but not pleasant. If I am correct, Mozart was too shrewd to let Beethoven pursue it.

In 1813 another friend enters the picture, Ferdinand Ries.

Ries was born in Bonn in November 1784. The Ries and Beethoven families were on friendly terms, it was the Ries family who nursed Beethoven's mother in her final illness in 1787, and Franz Anton, Ries's father, who was one of Beethoven's many early teachers. One of the best reasons to put Mozart in Bonn in 1792, one of the best justifications for Mozart taking Beethoven to Vienna, is what Ries did in 1799, when he was only 14: He abruptly left Bonn, first for Munich, but eventually for Vienna.

Like Beethoven, Ries was hugely talented but, unlike Beethoven, Ries was also ambitious. Bonn's court had closed in 1794, taking the court orchestra with it. There was no longer any musical life in the city on the Rhine, there was therefore no longer any reason for talented children to remain there. Age 14, Ries was still years away from puberty. He walked to his destination. While on the road he contracted smallpox. It blinded him in the right eye.

What would account for this child leaving home at such an early age? Without work, the Ries family was poor. Ries stopped off in Munich, in part to study with a teacher (to whom he eventually gave instruction instead), while copying music for a penny a page to support himself. With him Ries carried a letter of introduction to Beethoven, but, as we can see, and as the Ries family well-knew, Beethoven was expected to introduce Ries to Mozart.

Mozart's removal of Beethoven to Vienna was a red letter day in Bonn, and long remembered. Beethoven's subsequent success in Vienna was noted in his home town. Ries was therefore eager to go and join him and did so at the earliest opportunity. In other words, Ries did not go to Vienna to meet Beethoven, but rather, Mozart. Regrettably, secondary evidence suggests that Mozart and Ries did not get on. Ries left Vienna.

After years spent travelling on the continent and several close calls with the French army under Napoleon, Ries arrived in London in 1813, where he quickly came to the attention of Johann Peter Salomon, himself a native of Bonn, who introduced him to the newly founded Philharmonic Society, which is still in existence. Ries remembered his old friend in Vienna and acted as his London emissary, selling his works and arranging performances, many of which Ries personally conducted.

For his encouragement Ries got in return only piano sonatas and second rate overtures (Ruins of Athens, King Stephen, Wellington's Victory) from Beethoven. Which was journeyman work at best. Ries himself arranged the commission of the 9th in 1817, expecting to have a work within a year, but it was not finished until 1825. Ries gave an early performance, on May 23, 1825, at Aix-la-Chapelle. The score, direct from Beethoven himself, arrived so late that a complete set of parts could not be had (all copied by hand), so the performance omitted the Scherzo and part of the Adagio. There were a total of 432 players. I mention this as Sir George Grove mentioned it in his book, *Beethoven and His Nine Symphonies*. For most things, there is a simple explantion, which is the case here.

It seems that after 1812 Beethoven fell back into an insularity from which he was unable to escape. I once asked a clairvoyant why Beethoven went deaf. The answer was, *Because he would not listen.* I think this is true. In the chart, note the debilitated Mercury squeezed between Sun and Moon, opposed by Mars.

You might conceptualize this as Mercury's eagerness to hear (Sagittarius is always eager), but, trapped between a very loud Moon and a very loud Sun (tight conjunctions) and opposed by a very noisy Mars in Gemini, it's a wonder Beethoven could hear at all. He complained of a buzzing in his ears.

Beethoven is an interesting example of how fans can unwittingly "take on" and "become" their object of adoration. During Beethoven's lifetime there were dozens of other composers, working not only in Vienna, but also in London, Paris, Milan, St. Petersburg, Madrid, Berlin and numerous other cities. During his lifetime Rossini, not Beethoven, was regarded as the greatest living composer. Muzio Clementi was regarded as equal or greater than Beethoven. Clementi's sonatas are said to have inspired Beethoven's late sonatas. Among Beethoven's contemporaries are Papa Haydn, Franz Schubert, Johann Hummel, Antonio Salieri, Carl Maria von Weber, Mozart (in disguise), Ries (and Ries' great rival, John Field), Boccherini, Louis Spohr, Jean-Baptiste Cramer and many others.

The years from 1790 to 1830, the "Beethoven period" was one of the richest in all music, but the greatness of Beethoven, combined with his intense insularity, have obscured most of his peers and trivialized the remainder. We are only just now, in the last twenty years, coming to a proper understanding of this period. The coming decades will bring many surprises.

To conclude, and as you have asked, and as I happen to know, here are the programs to two of the Symphonies:

Symphony no. 3 in E-flat, op. 55, "*Eroica.*" The first movement describes the chaos and excitement of the French revolution itself. One *citoyen* after another jumps to the fore, the excitement is palpable. Do you hear Napoleon's footsteps? No, you do not. You hear the King's spies, *beware!* Beethoven uses short phrases and ever-changing orchestration to graphically show the tumult. Unique to this movement, in the development and coda, Beethoven varies his concepts, rather than actual musical motifs. Are the two opening chords the guillotine which chops off the King and Queen's heads? I do not know.

The second movement, the famous *Marcia funebre,* is built upon the sounds a man makes as he involuntarily bursts into tears. Take the opening motif, suck in your breath, and then let the tears flow. Beethoven has captured this precisely, as any man may prove. The movement starts in the traditional minore/maggiore (minor/major) funeral form, but soon departs into straightforward fantasy/program music. The death mourned is that of Marie Antoinette, youngest child of Maria Theresa. The music explicitly describes the flight from Paris, the arrest, the trial, and the execution. Remember that Beethoven is not Berlioz. Then in a passage that is so painful it seems as if the mind has snapped, we hear the echo of Marie's head bouncing, followed by that of her son, the Dauphine, and the despair is absolute. The very last scene is the funeral and farewell which Vienna was denied. *Knowing the program*, this is some of the most intense music ever written.

The old order having been swept away, the third movement, the Scherzo, opens with Napoleon, representing the new. Note the shift, from explicitly programmatic music in the first two movements, to abstract or pure music in the last two. In 1804, when Beethoven wrote this symphony, he, like most everyone in Vienna, knew nothing whatever about Napoleon. Bonaparte selected the bee as an emblem of his rule. The busy bee, the worker bee, the organized hive of bees, all working under the overall direction of the queen. Beethoven liked this and used it as his motif. Right at the very beginning, the strings describe the buzzing of a hive. In the Trio, which is the center of the Scherzo, Beethoven gives us the center of the hive, where we find the queen, or in this case, Napoleon, giving wise rule. The program of the trio has long been surmised.

The fourth movement describes The Life of the Hero and is mythical in nature. It is in the form of theme and variations. We do not start with the theme itself, but with its bass line, its harmonics. In myth, we do not start with the individual, but with his ancestry. So it is here. Simple variations are succeeded by ever more complex, more grandiose variations, until we reach an apex, at which time the hero falls ill, dies and is gone. His people are bereft and grieve. When, in 1819 (if memory serves) Beethoven was reminded of Napoleon and the *Eroica* he remarked that he had already written Napoleon's funeral music. Beethoven was referring to this passage, not the funeral movement itself.

But, in reality, the Hero never dies, he is merely in hiding. When the call is sounded (the trumpets) the Hero returns. This is the finale to the symphony, the triumphant return of the Hero, who visibly strides about, to great acclaim. Note how the music pauses briefly to permit a clarinet arpeggio, as if to say that everyone will be included.

Overall, the symphony shows the great care that Beethoven took with his music, in its conceptualization and in devising the themes and musical forms to fit.

Symphony no. 9, in d minor, op. 125, the "Choral." The work is misleadingly titled. This is the Karl symphony, the first three movements graphically describe Beethoven's relationship with Beethoven, Ludwig with Karl. A great deal of care was given to the opening bars. They describe a mystery that quickly becomes unpleasant. The first movement describes one Beethoven as the unstoppable force, and the other as the immoveable object. I regret I cannot assign one to Karl and the other to his uncle. The balancing of irreconcilable forces is the reason why the movement often goes in two different directions at the same time, without resolving either of them. A century ago Gustav Mahler wanted to improve the symphony by reorchestrating it to include modern forces, but his effort was botched, as he upset the very careful balance between the instruments. The first movement is complex and has never been understood. With few exceptions, conductors merely try to get through it.

The heart of the first movement — the heart of the symphony as a whole — is the beginning of the recapitulation. It is as clear a contrast to the original opening as is possible, but conductors refuse to play it as written. To his formless, rhythmless original opening phrase, Beethoven has imposed a most severe, most drastic beat, shown explicitly in the tympani. The trip-hammer theme itself is now *in contrast* to this. Where it originally floated, disconnected from all reality, in the recapitulation it *trips* and *falls* and *staggers* uncontrollably. Once this is realized, the enormous rage and cruelty of the long notes in the winds and horns and drums are understood. The passage as a whole evaporates, showing a mind under severe stress, it simply snaps.

The second movement, which is taken as a joke, shows Ludwig and Karl out visiting. Beethoven inverts 2nd and 3rd movements as he has still more to say about Karl and himself. Here, Ludwig is carrying on a conversation with a friend, and Karl is sniping at him. At length Ludwig turns towards Karl and screams, and then, with a big smile on his face, turns back to his friend as if nothing has happened. Later there is an uncontrolled fight.

The third movement was the hardest to puzzle out. You are wondering why I believe the 9th to be about Karl. Beethoven himself gave me the key to the first movement, in an extraordinary and brief encounter when I was 19. It took 30 years to explain it to myself, there is no hope in explaining to others. I used it to quickly unlocked the second movement, but the third had to wait until I heard Karl Böhm and the Vienna Philharmonic, who are, to my knowledge, the only performers to understand this cloying and pleading movement. In it, Ludwig tries to reach Karl directly, tries to explain that he is harsh, but only for Karl's benefit. Karl listens in mute rage, says whatever he needs to say to get away from his uncle, and then promptly acts out. Which sets up a replay, leading to an ending where a false hope has been stuck on very real despair.

The fourth movement, the choral finale, is a complete mess. Beethoven is trying to achieve two goals at the same time. On the one hand he wants to resolve the first three movements, on the other, he wants to set Schiller's Ode to music, a project he has been thinking about for more than 20 years. Not having written a symphony in five years (as of 1817, the date of the commission) and in increasingly poor health, Beethoven is not certain he will get another chance. It's now or never.

At length he put the symphony aside and wrote his second mass, the Missa, which runs nearly 80 minutes and which is one of the supreme masterpieces of choral music. It is his finest composition. The choral writing is exquisite. It was written for the investiture of Beethoven's patron, the Archduke Rudolf, but was not finished until years later. Composed immediately before the choral finale of the 9th, Beethoven displays the most masterful handling of orchestra, chorus and soloists.

All of which are lacking in the finale of the 9th. Those who consider this to be a great work are ignorant of Beethoven's abilities. The final movement opens with a search for a theme. Themes from each of the three earlier movements are considered, and rejected.

Beethoven finally settles on what appears to be a variation on the English song, *Yankee Doodle Dandy*. This was observed as far back as the 1860's, in Boston (if memory serves). In previous years Beethoven had set some 179 English and Scottish folk songs, a part of his work that is largely unknown today. It is therefore reasonable to believe that he knew of *Yankee Doodle*, and that he knew it was vulgar to English tastes (remember it was the London Philharmonic Society who commissioned the work). Beethoven's use of it in a symphony that is stressful and difficult, graphically shows his mood at the time.

Beethoven wrote the finale of the 9th in a rush, off the top of his head, which shows how great a composer he was, that a casual scribble was of such merit. Many of the sections, including the *presto* and *prestissimo* passages at the very end, are *pro forma* and of no great merit. I could say more, but it would be personal opinion, and there has been enough of that.

This essay was first published in the April 3, 2012 newsletter. It has been revised and expanded.

Now it's back to astrology. — *July 24, 2012*

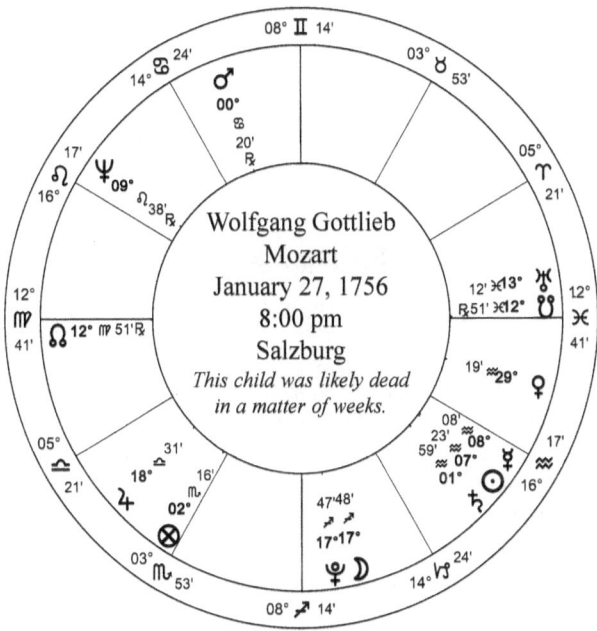

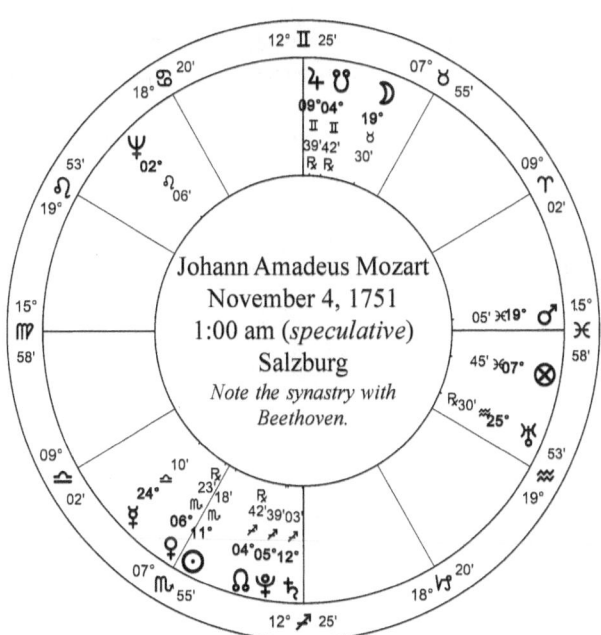

🕐 Spring Planting

Is not just yet. I was sorely tempted to plant in the March heat wave, but kept telling myself that April is a colder month, and, indeed, it has been. Even if it's still warm where you are, you don't want to plant in the dark of the moon (full to new) as plants will get a weak start. And you do not plant on the full moon: Plants will shoot up quickly and then fall over.

For general garden crops, Friday April 27 will do nicely: Moon in Cancer.

Flowers do nicely with a Venus sign, which is April 21-22 (Moon in Taurus), but as that's the new moon, they will get a slow start.

Root crops—potatoes, beets, onions, carrots, etc., are planted in the dark of the moon, since you want root development. Try the Moon in Pisces, April 16-17. Avoid Moon in earth signs, as that promotes woodiness.

Weeding. Weed your garden in the dark of the moon under a fire sign. Weeds will be a long time coming back: April 19-20, Moon in Aries, about to go new.

It's a bit early to think about harvesting, but that's under a waning moon in air signs. Air signs promote dryness and avoid rot. —*April 10, 2012*

Know Your Leaders
🕒 ANTONIN SCALIA

ANTONIN SCALIA, current Supreme Court Justice, was born on March 11, 1936, in Trenton, NJ, at 8:55 pm, birth certificate in hand. How is it, you ask, that a Supreme Court Justice has put his birth certificate up for inspection? Well, he didn't. When Ronald Reagan nominated Scalia for the Supreme Court in 1986, the late Lois Rodden phoned her contact in Trenton and got the birth time.

For many years Rodden ran a data service (Data News) which gave actual recorded times of birth for virtually every celebrity in the news, as soon as they popped up into view. She never said how she got the times, only that she somehow got them. I really, really miss Lois, her spunk and daring. Like as not her birthtime rating system will eventually be abandoned simply because so few real birth certificates will come to hand, while, on the other hand, there will be (and has already been) much abuse.

Scalia is now 76 years old and is the senior serving member of the current Supreme Court, having been appointed to the court in 1986. You are wondering how long it will be until he retires, or heck, just plain passes away and are secretly hoping, perhaps, that I might use old astrology to tell you.

And that would be a wicked, wicked, evil thing to do. Because even if it's just a matter of conjecture and even if every case I've presented to you, so far, has been more-or-less right (those scheduled to die young did, in fact, die young, even if the prediction was wrong by five years), predicting the date of a living person's death is a horrible, cruel thing to do, and it's a horrible and cruel thing even if Mr. Justice Scalia is a public figure of long standing and even if I am an obscure workaday hack that no one has ever heard of. It's just not done, okay?

On the other hand, those who, by birth or by merit or by sheer accident arrive at power and are responsible for the public good have the obligation to exercise that authority wisely, and for the good of the population in general. Such people are required to not only be *impartial,* but in fact to be *generous.* A head of a corporation can be greedy and self-centered and stupid, since he is merely the representative of his company. **A public official represents the nation's people as a whole** and must always keep this foremost in mind. Ideology may be tolerated, but only in small doses, and with incessant apologies.

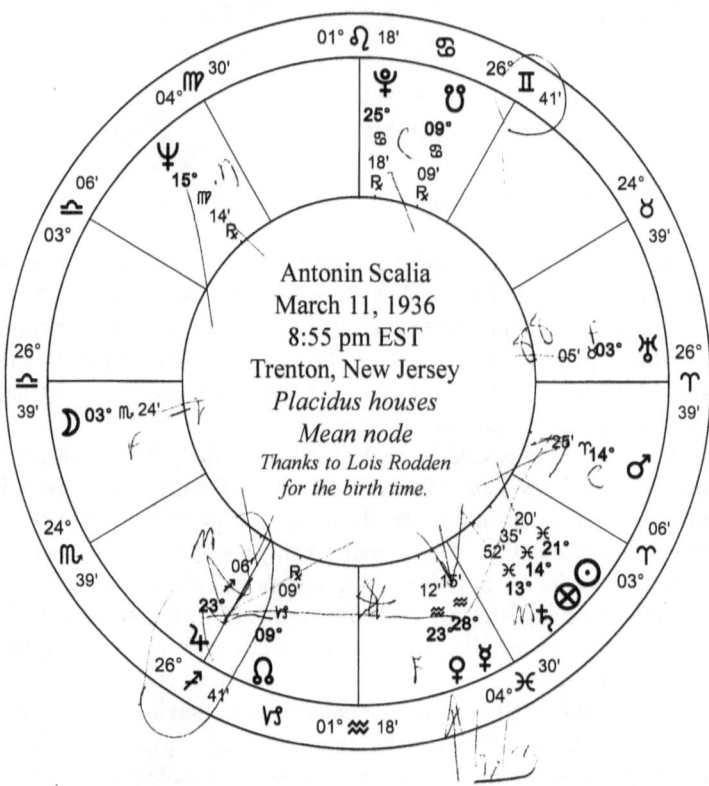

But it seems to me the government of the United States has forgotten this essential rule and now governs in defiance of its people. A self-aware astrologer is one of the most powerful of all people and he exercises that power when he reminds his leaders of their own mortality. In other words, it is we, as astrologers, who are the fabled Fourth Estate of government. The press might have been co-opted. Astrologers cannot be. There's just too damn many of us.

Looking at his natal chart, you would not think Antonin Scalia to be inherently powerful, nor inherently brilliant. Libra rising, ruler Venus conjunct Mercury in Aquarius, you would think of him as an affable storyteller, and, indeed, Scalia is. Libra rising, Venus conjunct Mercury is an aspect of grace as well as eternal youth. Scalia, in fact, appears far younger than his actual age. A Venusian touch in a natal chart is always an advan-

Antonin Scalia 137

tage, one that Bill Clinton, Libra rising, Venus in the first house and in Libra as well, knows well. It smooths one's path in life.

Intellectually, Scalia's chart is upside down and we are mistaking cleverness for actual genius. This is where it helps to have a biography as a guide. Genuine cold readings are quite hard to do, as I have learned to my dismay.

In Scalia's chart, Jupiter is in Sagittarius — which it rules — and on the third house cusp. The third house has to do with details, but with Jupiter here we cannot be bothered with them. Details are not important. Concepts are.

Since I myself am concept-driven I would support this, but *details must prove concepts*. This step cannot be omitted, concepts may not stand on their own. Ideally, the concept in question must be so good and so secure that not only will *no detail, nor any collection of details, countermand it*, but the concept will be so powerful that it goes on to *organize previously unrelated details*. Such as my personal theory that the signs of the zodiac are not in the sky but are qualities inherent in the Earth itself, but I digress.

In Scalia's case, we then look across the chart, to the 9th house, the House of Concepts, where, of course, we find Gemini. Gemini is as out-of-place on the 9th as Sagittarius is on the 3rd. Now everything hinges on the condition of Mercury in the chart. Since Jupiter is powerful, it will throw up an unlimited number of grand ideas. If Mercury is in, say, Virgo in the 11th, then its rulership of the 9th will tend to shoot down a lot of the Jovian hot air that will come across the chart to challenge it. The square from Mercury to Jupiter itself will make the process stressful, which will tend to make the native eternally uncertain of himself. Make him "try harder" and do better.

This imaginary Mercury in Virgo in 11 would also be opposed to Scalia's Sun-Saturn conjunction in Pisces, and, as it would tend to actualize the 11th house itself, Scalia's friends would be eternally carping at him. (Nevermind that Sun-Mercury oppositions are impossible.)

Instead, in Antonin Scalia's chart, Mercury is in Aquarius, a sign which it quite likes, and which lets it focus on large, social issues. While Mercury and Jupiter are natural enemies (they rule opposite signs), in Scalia's chart—as in mine, for that matter—they are friends. They are not only in signs that each of them like, they are also sextile to each other. Jupiter, in Mercury's house but in Jupiter's sign, is ideally placed to tell Mercury how the third house should be run, and even though Jupiter will be wrong, Mercury, strong by sign, sextile by aspect, and having no real interest in the 9th (which it rules), will let it.

Nor will that ever trouble Mercury very much, since, as you can see, Mercury is in trine to the cusp of the 9th itself. It's not just that trines are aspects of ease and grace, they are also aspects of *benign neglect*. In this regard, Venus's conjunction with Mercury merely reinforces this rather enormous blind spot. Antonin Scalia does not try to get the details correct because he doesn't really have to. Concepts are what he's got, and concepts are all that are important.

The nature and quality of these concepts are shown by aspects to Jupiter. In addition to its sextiles to Mercury and Venus, Jupiter has many other aspects.

Jupiter is square to the Sun: Optimistic, buoyant, generous, but wasteful.

Jupiter is square to Saturn: Restless and seeking. Note all three planets are in mutable signs, they are inherently unstable.

Jupiter is trine to Mars: A true believer.

Jupiter is square to Neptune: Illusions, dreams, unreality.

Jupiter is inconjunct to Pluto: As inconjuncts are aspects of invisibility and as Jupiter is expansive and Pluto is powerful, the inconjunction is not being aware of his own power, of his own strength. Overbearing, in other words. In fact, Scalia has always been a dynamo.

Having a fundamental grasp of the man, the rest of the chart is simply sketched.

Moon in Scorpio is an intense desire for emotional control, of himself and others. It is the ability to chop others off, which Scalia has in abundance. Obscurely placed in the first house, others are shocked when such an urbane and friendly man (ruling planet Venus conjunct Mercury in Aquarius) suddenly adopts crude behavior, such as his famous hand under the chin cupping gesture, a common gesture in Italy for expressing indifference, and which Scalia has used to brush off critics. It is not technically obscene, but it is a clear *let 'em eat cake* gesture. The Moon's opposition to Uranus is an antisocial craving for excitement, which merely reinforces a Moon in Scorpio's antisocial nature.

Sun-Saturn conjunction, which is both wide and waning. This always results in a serious attitude, which, in Scalia's case, both wide and waning, he would like to dispense with. Aspects are dynamic in this fashion. Both planets in the 5th house, he would be a stern father. He has nine children, five boys and four girls. Why so many? While I am not skillful enough to actually count them in the chart, note that both Sun and Moon are in fertile water signs, a fertile sign is on the 5th cusp, which is ruled by Jupiter which is itself in its own sign, and even if Sagittarius itself is not fertile, Jupiter's many aspects certainly enhance its ability to deliver kids. Saturn in the 5th means that one child may have been lost, which can be by miscarriage. Scalia was 24 when he married, which, judging by his chart, was late. (As a rule, the more fertile the chart, the earlier the marriage.) Even though the year was 1960 and the sexual liberation was still

Antonin Scalia

in the future, I doubt he was a virgin on his wedding day.

Scalia's **Sun is opposite to Neptune,** which is an aspect of deceit (deceiving others, and they deceiving him, note especially Neptune's placement in the house of would-be friends), but in Scalia's chart, the Sun-Neptune opposition is overwhelmed by the **Saturn-Neptune opposition,** which is much tighter. Alan Oken describes this as "swimming with leaden boots," which is to say, trying to get by without paying your dues, and acting irresponsibly in general.

Note **Mars in Aries in the 6th.** Mars in the 6th is generally a hot-headed troublemaker at work. In this respect, Mars in Aries in the 6th is much easier to deal with than, say, Mars in Scorpio in the 6th, which tends to be underhanded and devious. Mars in the 6th gives inexhaustible energy for work. And, in fact, during his time on the court, Scalia has written scads of opinions, majority opinions, concurring opinions, dissenting opinions. Unlike some jurists, who cultivate or manipulate other jurists to sway them to their opinion, Scalia charges straight ahead, fearlessly, recklessly, a ram butting his head.

Lastly, Scalia's **Sun in Pisces.** Weakly placed in Pisces, it is square to Jupiter, both in mutable signs, and trine to Pluto, which gives Scalia's Sun what little positive strength it has.

It is said that Scalia has always been conservative, even when he was a student in the 1950's. There is nothing in his chart that would make him a radical, whereas his Catholic upbringing, with its emphasis on tradition and authority, would tend to work in the opposite direction. The Second Vatican Council, which ended in 1965, was very likely a turning point, as it was for me and virtually all Catholics alive at that time.

In Scalia's case, the revolutionary change in his fundamental belief system came during the height of his first Saturn return. The Church's long-overdue reform became forever tied to Scalia's transition from late childhood, to full-fledged adult. Like many of us, he repudiated the Church's change. In his case, he became a member of the breakaway Tridentine faction, which continued with the Latin Mass.

Note carefully how this plays with the various factions in his chart: Sun, Saturn, Jupiter, Neptune in mutable signs: The world is a crazy place that will do stupid things. Moon-Uranus-Mercury-Venus in fixed signs: We must be strong and stay firm. Mars and Pluto in cardinal: Maybe if we work hard we can find the power to change things.

Or, more simply:
The *Sun* says we are weak and at risk.
The *Moon* says we must remain firm.
Maybe we can find the power.
Such is the chart of Antonin Scalia.

Having had his religion shot out from under him in his youth, Scalia will not let this happen again. Challenges to tradition will be suppressed. Remember that for Antonin Scalia, details are not important. Concepts are.

We now understand why this man has been at the forefront of innumerable idiotic Supreme Court decisions. To cite a few of them:

Bush v. Gore, 2000, in which the Supreme Court forcibly stopped the State of Florida's recount. The Constitution clearly gives the States, and the States alone, the power to count votes.

Citizens United, 2010, which gave the rich unlimited right to buy elections.

Florence vs. Board of Chosen Freeholders of the County of Burlington, 2012, which gave police the right to strip search anyone, anywhere, at any time, because the alleged risks to law enforcement outweigh all other considerations. Think the cops won't use it? This is beyond repulsive.

Antonin Scalia was part of the majority of each, and as the longest serving justice, has had a major role in shaping the current court overall.

Last fall's Occupy movement proved that peaceable protest will neither be heard, nor even tolerated. Last Tuesday was primary day here in Maryland. I was so angry I could not make myself go and vote.

I am an astrologer and will therefore use the tools at my disposal. **Here is how we calculate Justice Antonin Gregory Scalia's life expectancy:—**

We must first determine the *hyleg,* or giver of life. In a daytime chart the Sun is normally the hyleg, but Scalia was born at night.

We next consider the Moon. For a planet to be hyleg, it must be in aspect to one of the planets that rule the degree in which it is located, which, for the Moon, is 3 degrees Scorpio. Those ruling planets are: Mars (sign), Mars (triplicity/night), Mars (term), and Mars (face). There is no exaltation. The Moon is not in Ptolemaic aspect to Mars and therefore is not hyleg.

We next move to the ascendant, Libra. As the ascendant is a point and does not "casts rays" we do not consider it, but rather, its ruler, Venus. Venus is at 23 Aquarius. Saturn rules the sign, Mercury rules the night time triplicity, Venus rules the term, and the Moon rules the face. As Venus is conjunct Mercury, this makes Venus the hyleg.

As the *alcochoden,* or giver of *years,* is defined as a planet which both rules and is in aspect to the hyleg, this makes **Mercury the alcochoden.**

The *alcochoden* gives *years* based on the *house* in which it is found. Angular houses (1, 4, 7, 10) give the greatest years, followed by succeedent (2, 5, 8, 11) and then cadent (3, 6, 9, 12). Here is the table, so that you may know:

Antonin Scalia

	Angular	Succeedent	Cadent
Saturn	57	43½	30
Jupiter	79	45½	12
Mars	66	40½	15
Sun	120	69½	19
Venus	82	45	8
Mercury	76	48	20
Moon	108	66½	25

These numbers are not arbitrary, but based on an actual, calculated sequence, which, forgive me, I forget at the moment.

Mercury in the 4th house (angular) as alcochoden, or giver of years, gives 76, which is Scalia's current age. To this we are to add the minor period of Jupiter, if it is conjunct, trine or sextile. As Jupiter is not only sextile, but strongly placed in its own sign, we may confidently add its minor period of 12 years. We then do the same with Venus, which is conjunct the alcochoden. Venus's minor period is 8 years.

We must then subtract the minor periods of Saturn and Mars, if we should find them conjunct, square, or opposed to the alcochoden. As neither planet is, we find Justice Scalia may reasonably expect to live to his 96th year (76 + 12 + 8). Which will take him to the year 2031. He will be with us for a long time to come. If he is still a member of the Court, he will be in his 45th year there.

In messing around this week I did the same calculation for Justice Kennedy and discovered, to my surprise, that he may reasonably expect to live at least into his mid 80's. As both of these results are far beyond average, I was then led to the perhaps not surprising conclusion that longevity of one's forebears is a major consideration in the selection of a Supreme Court justice, and presumably always has been. You think that heredity is not part of astrology? Of course it is. We incarnate as groups. Astrology merely reflects the qualities of our group.

Those who rule a country are members of a small and exclusive club. In the case of those who rule the United States, not all of their birth times are known, but I am sorely tempted to publish the calculated dates of death for all those whose times are known. If I do so, it will not be because I wish them ill-will, nor do I wish to cause their families or friends distress. Fake elections, the suppression of dissent, and the shocking legalization of strip searches, are my own personal limits. It is time to gently remind our rulers of their intolerable, inexcusable misrule.

And then wish them long and happy and peaceful and prosperous lives. ✠ *— April 10, 2012*

☉ Earthquakes

This month there have been two notable earthquakes.

The first was a magnitude 7.2 in Baja, Mexico, on April 4, at 22:40 GMT at 32N, 115 W (data from the US Geologic Survey). The chart has 3° Virgo rising, with a MC of 0° Gemini.

The second was magnitude 8.6 in Sumatra on April 11, at 8:38 am GMT. According to the USGS, it was at 93° E, 2° N. It had 0° Virgo rising, with 4° Gemini on the MC. Both charts with very nearly identical angles, both charts with Mars on the ascendant and Neptune on the descendant, with the nodes in 4 and 10.

George McCormack, in his *Text-Book of Long Range Weather Forecasting*, says quakes are touched off as a result of MC/IC placements at the preceding equinox or solstice. Which would be the Spring Equinox, at 5:15 am GMT on March 20, 2012. For the Mexico quake, this gives 19° Leo on the MC. For the Sumatra quake, 19° Pisces MC. The Moon-Neptune conjunction opposite Mars falls in 4 and 10 in Baja, and 9 and 3 in Sumatra. I've not looked at a lot of quake charts so I am not quite sure what to make of it. — *April 17, 2012*

☉ My Dinner with Andrea

I had a delightful dinner on Saturday with Portland's Andrea Gehrz, who was in town to give a couple of lectures at the NCGR and Baltimore's Astrological Society. I showed her my progress on Vettius Valens (mostly done but not quite), we talked about what Greeks should be translated next. As Andy reads Greek fluently, including the New Testament, I asked her if its original verses would be easy to memorize. She said yes, quite easily memorized. Which confirms my hunch, that the early Church was an oral affair and the Gospels only put to paper when the Church was growing too large and too fast. It is still largely hidden. — *April 17, 2012*

Know Your Rulers
☽ **Joseph Biden**
Vice-President of the United States

Joseph Biden was born on November 20, 1942, at 8:30 am EWT, in Scranton, PA. The Rodden rating is A, which is to say, she had a report from someone who said he had seen the birth certificate, and in the years since, that report has not been contradicted. The 8:30 am time is therefore considered to be reliable.

Joseph Biden has 3 degrees of Sagittarius rising. The ruling planet is Jupiter in Cancer in the 8th. Jupiter is exalted in Cancer. I am looking at my textbooks, which say that Jupiter in the 8th house is mystical and spiritual and occult. And that you marry money and have inheritances, which for Joseph was true of his first marriage, but 8th house placements turn up too often in political charts for that to be the whole story.

I am coming to a different understanding of a politician's 8th: *What can I get from you that I can use for my own benefit?* Or more bluntly, *How can I pick your pocket?* Politicians always have the most charming way of doing that, which is why we elect the scoundrels. This is shown in Joe's chart by friendly, enthusiastic Sagittarius rising, with ruling planet Jupiter in caring and concerned Cancer in the 8th. Having gone retrograde a week before his birth, Biden's Jupiter will ultimately care more for himself than he does for you, sweetie.

I am getting mean about this because the economic situation in the US continues to worsen yet I do not see our leaders running wildly about the country with their hair on fire trying to do something about it. This is all the more bizarre in an election year, when our rulers usually pretend to be serving the populace. Instead they seem like Joe: Placid and serene. As if the upcoming election was, well, fixed, maybe.

In college, Biden was an indifferent student. Wiki says he was distracted by an out-of-state girlfriend. Can we see this in the chart? Yes. Third house of mundane studies has Aquarius on the cusp. (Joe said he was bored in school. Scholastic boredom is third house, not ninth.) The ruler of the third is Saturn, posted in the 7th house of girlfriends, and in

Gemini. So rather than study, Joe leaned on the ruler of the 3rd and found girls. Let's try a new wrinkle in chart reading. Let's say that not only does the ruler of the cusp bring the affairs of the house it is domiciled into the house it rules, let's also say it brings the sign it's in as well, and then let's look at how the "relocated sign" gets on in that house. Got that?

Okay. Saturn in Gemini in the 7th is ruling Joe's 3rd house. As we know, Gemini likes the third. Can this explain why Joe was stuck with out-of-state girlfriends? Would not Saturn in *Gemini* in the 7th make it hard for Joe to get *local* girls? Make it *easy* for him to travel?

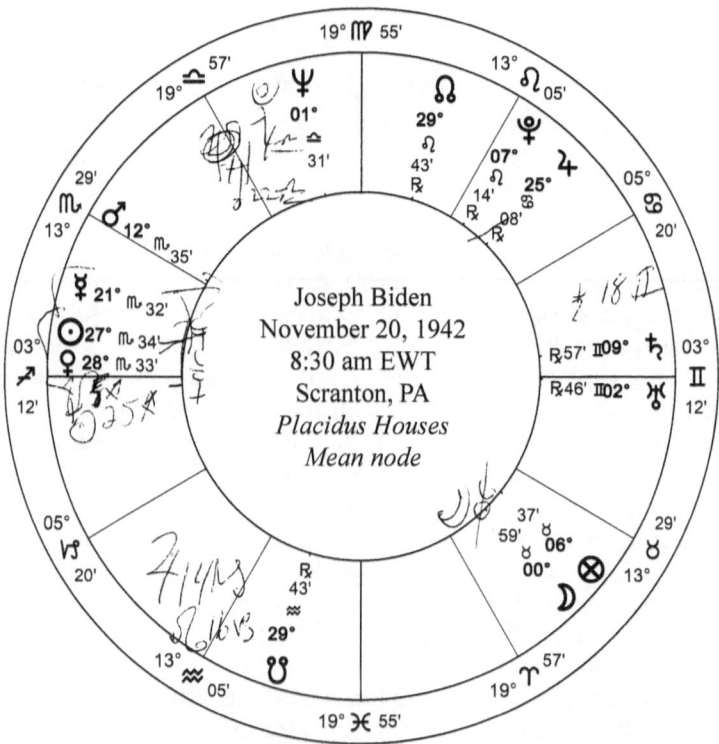

Let's look at it from the girl's point of view. Take Joe's 7th house as her first. Her first house Saturn rules her 8th house of sex and martial duties, and also her 9th house of travel and adventure and higher education. Saturn also has strong influence in her 5th house (Joe's 11th), with Libra on the cusp. And into her boring, Saturnine life, up pops safe and sedate college-student Joe. Exactly what she was looking for.

Remember that when you turn the chart like this, it DOES NOT describe the whole world. It describes the world FROM THE POINT OF VIEW OF

Joseph Biden

THE NATIVE. The women in question will only be "like this" when they're with Joe. The rest of the time they're themselves. Joe's chart is only "renting" them. Have you ever noticed how you can be "different" around different people? It's not just planet-to-planet synastry that's going on. You're also spinning each other's charts.

After college, Joe tried his hand at being an attorney, first as a partner in a firm, and later on his own, without success. In 1970, he ran successfully for City Council at New Castle, Delaware. At the age of 27, Joe had found himself. He was a success. He never looked back.

In 1972, Biden got into national politics by running for the Senate seat held by Republican J. Caleb Boggs, who was on the verge of retirement. Biden used his agreeable, friendly nature in a door-to-door campaign. Delaware is just that small, and though he trailed in the summer, he won in a squeaker.

Barely a month after that, on December 18, 1972, his wife and youngest daughter were killed, and his two sons gravely injured, in a traffic accident in Hockessin, Delaware.

It would seem that Mrs. Biden pulled into an intersection and was hit by an oncoming big rig. Tractor-trailer rigs, when fully loaded, can weigh upwards of 40 tons, they are neither nimble nor agile. As Wiki says the truck driver was cleared of wrongdoing, it would appear that Mrs. Biden was foolish. I once knew a truck driver who killed a family in precisely this way. He watched his truck roll straight over a car full of people, in speechless horror. It shattered him.

It shattered Joseph Biden, who had just turned 30, was the 6th youngest man ever elected to the Senate, and who had not yet even taken the oath of office. There is no greater tragedy than to lose one's family. That it happened at the very moment of his greatest triumph made it all the more intensely severe.

Let's look at his transits on the day:

Having just arrived at the age of 30, you will wonder about Saturn. It was at 18 Gemini, past the return at 10 Gemini, but still in the 7th house. One delineation of natal Saturn in the 7th is the eventual death of a spouse. A definition of transiting Saturn in the 7th is the end of a relationship. Biden's natal Saturn lacks significant Ptolemaic aspects (trine to Neptune, sextile to Pluto) and has, instead, only "invisible" inconjunct/whole sign aspects to Mars, Mercury, Sun and Venus. This might have been the very first time in his life that Joseph Biden had ever really experienced Saturn.

Transiting Mercury was at 4 Sag, conjunct the ascendant, and also conjunct transiting Neptune, at 5 Sag. Neptune on the ascendant, Joe had entered a world of make-believe. When Mercury transits the ascendant it brings news to the native. In Sagittarius the news will deal with hopes and ideals, and also transport and distance.

To drive this home, transiting Mars at 20 Scorpio, was conjunct Biden's natal Mercury at 21 Scorpio. Note the tendency for transiting Mars to "strike early," while transiting Saturn comes late, in this case, eight degrees "late."

Transiting Venus was directly conjunct Biden's Sun, the transiting Moon was in Taurus, where Biden's Moon is natally. On the day of the tragedy, Biden's ascendant, Sun, Moon, Mercury, Venus and Mars were not only all tightly transited, but all the transits were conjunctions. It was like being hit in the face with a dozen hammers.

In grief and shock he wanted to resign but was persuaded by Senate Majority Leader Mike Mansfield to take the oath of office, which he did at the bedside of one of his two surviving sons (Beau or Hunter), the other members of his family having died. It is believed that Biden never recovered from this loss.

Further note the placement of Mars in Biden's chart. Mars rules the 5th house of children and is itself in Scorpio in the 12th, which is the 8th house from the 5th, or, in other words, in the house of the death of children. Mars in Scorpio strikes by stealth. You will never see the blow coming, you will never recover from it.

Otherwise, Biden's chart, like the man himself, is not of great interest. Joe Biden's life demonstrates the curse of being a successful man, in that once one is successful, one stops trying. This is in part as he no longer needs to try, often compounded by the fear of losing what he has. The earlier the success, the more complete the ultimate failure.

Sag rising, Jupiter in Cancer, Mercury-Sun-Venus trine to Jupiter, Joe is an affable man. You will note mutables on the angles, which indicates a life that never quite settles down. Angular rulers, Mercury and Jupiter, are in trine from 12 to 8, the two most "hidden" houses in the chart. Joe Biden's long career in the Senate was undistinguished, save for his sponsorship of Amtrak. Why was this? Because his home in Delaware was close enough for him to commute to Washington on a daily basis. Joe went by train.

From his ascendant, we can already see Joe's maladroit speaking style, his habit of putting an appendage in his mouth and chowing down. With Sag rising, we expect youthful boyishness and enthusiasm, with Jupiter the ruler in Cancer, we would expect heartfelt caring, in the transcendental 8th we expect an intense, intimate point of view, but, the ruler retrograde, it all gets messed up. Retrogrades are like dull knives. They never work the way you expect, and if you're not careful, you can blunder

Joseph Biden 147

and hurt yourself. This is especially true when the chart ruler itself is what's retrograde. While head of the Judiciary Committee, Joe was known to ramble on so long that no one could remember what his point was. His natal Mercury is in Scorpio, where it hears intensely, but in the 12th, hears mostly itself. Which, by the way, is as bad a placement for a politician as it is for a musician.

As Vice-President, Joe Biden joins a long line of forgettable people. The American system rarely puts brilliant men in this position. The Vice-Presidency has often been a dumping ground for politicians from small states, states where one can still campaign door to door and actually meet the bulk of real voters. Cheerful, sunny glad-handers are often elected by this means. Misled by easy victories, they are frustrated in their efforts at higher office as they are rapidly—and easily—chewed up by the pros who know what they're doing. The lucky ones (Dan Quayle, Spiro Agnew, Walter Mondale, etc.) may find themselves selected for largely honorary post of Vice President, mostly because of their inoffensive fecklessness.

Longevity. Why not? It's a useful technique but you won't learn unless I drill it into you. Although a day birth, Biden's Sun is not hyleg, as the Sun is never hyleg when in the 12th or 8th.

Nor is the Moon, as it makes no aspect to Venus or Mercury, its relevant rulers.

Which brings us to the ascendant, ruled by Jupiter. Jupiter at 25 Cancer is ruled by a term (or bound) of Venus, with which it is in trine. This makes Jupiter hyleg and Venus alcochoden, or giver of years. I am going to fudge exact degrees, as I suspect the ancients could not reliably determine the ascendant to a precise degree, and declare Venus to be angular. An angular Venus gives 82 years. The trine to Jupiter gives him Jupiter's minor years of 12. There is no aspect to Saturn but there is a conjunction (whole sign) with Mars, which would deduct Mars' minor cycle of 15.

This makes for a net of 79. *Presuming ancient theory is correct,* **this is not the age at death,** merely the theoretical opening of a "window" which thereafter makes the native susceptible to a fatal transit of one sort or another. Theory says that *prior* to this moment the native will successfully survive all transits, directions and progressions, whatever they may be. You will note that Steve Jobs (*Duels At Dawn*) and Whitney Houston both lived a number of years beyond what ancient theory predicted.

The best longevity instructions (which are still far from perfectly clear), along with the longevity table, are given in *Judgements of Nativities*, by Abu'Ali Al-Khayyat, but the rulership table you need is missing. You will find that in both volumes of Lilly, and also in Al Biruni (fragmentary), Dorotheus and most good books on horary. Take a look at your bookshelf. You might already have it.

Traditional textbooks say that Jupiter in the 8th gives a peaceful death, and peaceful deaths are very often at a very advanced age. It seems clear that Joseph Biden will outlast his current job, which he must vacate by January 2017 at the latest. Will he go back to the Senate? I wonder.
— *April 17, 2012*

General Influence of the Planets when Stationary, or in the Equator or Tropics
From Long Range Weather Forecasting, *by George McCormack*

Neptune usually conduces to variable "freak" weather, lowering barometer, southerly winds, humidity, excessive static, hazes, fogs, sudden changes, more effectively in lowlands and along waterways. Combined with Mars, this planet is related to seismic phenomena. *Spring:* Misty and mild. Fogs at night. Sudden changes. *Summer:* Sultry; warm. If thick haze, sudden showers. *Autumn:* Lower barometer. Mild; misty. Showers at night. *Winter:* Damp and foggy. Vertical ascending currents. Unsettled.

Keynote of URANUS is high barometric pressure, descending vertical air currents, increased wind velocity and sudden changes to colder. Winds are gusty. Sudden frosts and cold snaps. Affects highlands first, then downslope. *Spring:* Overcast, cold and blustery. Chilly drizzles. Frosts at night. *Summer:* Winds shifting to N.W. Storms originate in highlands then south. Temperature falls. *Autumn:* Gusty. Chilly. Bleak skies. Fine rain. Often cold drizzle. Frosts. *Winter:* Windy and stormy. Fresh to strong N.W. winds. Cold wave follows.

Saturn's keynote is lowering barometer, steady but decisively. Shadows, dampness and cold. Easterly winds. Effects very general over large areas. Low hanging clouds. Excessive humidity, slow build up of low pressure areas with increasing cloudiness. Downfall under this planet is more lasting than with any other. *Spring:* Increasing cloudiness. Damp and wet. Colder. *Summer.* Overcast, humid, showery then colder. *Autumn:* Low clouds. North Easters. Rain and colder. From *Long Range Weather Forecasting,* by George McCormack. — *April 24, 2012*

The Foreword from Long Range Weather Forecasting:
☉ Astrology Under Our Feet
by David R. Roell

McCormack's book is finished, it went to the printer on Sunday. A proof will be back the end of the week and presuming I don't faint, the first stock will be here about a week after that. In other words, McCormack arrives in about two weeks.

In its intensity and drive it is one of the finest astrology books I have ever seen. I wrote the following as an introduction:

I was working on a theory of astrology when, in January 2011, Parke Kunkle, of the Minnesota Planetarium Society, casually repeated an old story, that there were really 13 signs of the zodiac, not twelve. Thus inspired, I redoubled my efforts. As long as it has existed, Astrology has been a puzzle with multiple unknowns, hence the real difficulty in making sense of it. I was further inspired by George McCormack's extraordinary book, the one you now hold in your hands. His observations fit my theory so closely that I dare to venture an overview of my work here, in the hopes that my theory and McCormack's observations may combine in some useful way.

The Tropical Zodiac

McCormack uses the Tropical Zodiac in his work. The tropical zodiac is based on the Earth-Sun relationship, specifically, declinations and seasons. Aries starts the spring season, Cancer the summer, Libra autumn, and Capricorn, winter. These are calculated as of the moment the Sun appears to cross the Earth's equator, or achieves maximum north or south declination.

The length of each season is based on the dates of the Earth's apogee and perigee, the Earth's closest and furthest approaches to the Sun. As these dates slowly drift over time, the exact length of each season, in days, hours, minutes, and seconds, slowly change over the centuries. While the length of the year remains exactly the same, its seasonal divisions do not. The four seasons are not of exactly equal length. Most likely, they never have been. In 2012, the Earth was at perihelion on January 5, and at aphelion on July 5. In 1837, these dates were January 2 and July 1. *(Source: Tables of Planetary Phenomena. 3rd edition, 2007, by Michelsen and Pottenger)*

The start of the four seasons are represented by the four elements: **Fire**, for spring; **water**, for summer; **air**, for autumn; and **earth** for winter. Each of the seasons were then trisected to show the three states of energy they embodied: **Active**, or Cardinal; followed by **Immoveable**, or Fixed; and then, **No energy**, which we know as Mutable. Whereupon the next season arrived to restart the energy matrix. An element was then assigned to each third of a season, resulting in a year of four seasons and twelve segments (months), which became the signs of the Zodiac. This is a precise method, and an unvarying one. The Zodiac is thus a 4 x 3 grid, of energies mapped against elements. There are only twelve signs. There could only ever be twelve. Not thirteen.

There is an old story that Libra and Scorpio were once joined as a single sign. This was presumably an attempt to suppress Scorpio as an "evil" sign. The inherent structure of the Zodiac makes eleven signs impossible.

This structure means the Tropical Zodiac is not *of the* sky, nor *in* the sky, but **expresses the relationship between Earth and Sun.**

Can we determine if the Zodiac is geocentric or heliocentric?

Yes, we can. First, we note that as we are talking of ourselves, the Zodiac we create will presumably be created of ourselves. Of the Earth, in other words. A plant, for example, grows *from* the Earth, but grows *because* of the Sun's light and heat. So is the plant of the Earth, or is it of the Sun? The answer is that the plant is *of* the Earth, but is *nurtured by* the Sun. The plant is therefore a subset of the Earth. It is the Earth's response to the Sun's energy. In its own unique way, the plant shows the Earth-Sun relationship.

We do the same analysis with the elements and energies that make up the Zodiac. Where can they properly be found? The Earth contains all four elements, fire, earth, air and water. The Sun has only fire. The Earth also has all three states of energy: Cardinal, Fixed and none. The Sun has only Cardinal. By comparison, the sky has neither elements nor energies. Therefore,

The signs of the Zodiac are qualities inherent in the Earth itself.

Astrology is Earth-based.

The Tropical Zodiac has nothing to do with the sky. Never has, and never will. Presumably the Zodiac was projected into the sky at some point as a reference, as in, *See this constellation overhead at midnight? When the Sun gets to it, it will be summer again.* (The Sidereal Zodiac has a different explanation. It is not based on stars, but the Moon.)

If the Zodiac is not of the sky but are qualities inherent in the Earth itself, then it is reasonable to expect the Earth would express these twelve basic qualities. Indeed, the word *zodiac* itself relates to animals. Traditionally it is said that Aries is symbolized by the ram, that Taurus is symbolized by the ox, etc., but it is more penetrating to say the Earth **expresses** its Aries energy *as* a ram, its Taurus energy as an ox, its Cancer energy as a crab, its Leo energy as a lion, etc. If qualities are inherent in the clay of the Earth itself, then it is natural the Earth would express them. By contrast, if astrological energies were external to the Earth, then the Earth's response would be *in reaction* to them. Sometimes agreeable and compatible, sometimes hostile and maladroit.

This analysis can be taken further. Medical astrology holds that different parts of the body are ruled by different signs. If the ruling signs are external to the Earth, then the Earth's reaction to them will produce maladjusted, maladroit results. Only if these qualities are *inherent in the Earth,* can the Earth produce perfect results, since those results will be in keeping with the fundamental qualities of the Earth and will be as perfect as the Earth, no more and no less.

In other words, the eyes, which in all creatures are found in pairs, are ruled by the Sun and Moon, the two light-givers, as represented by Leo and Cancer. Because it is the Earth which produces Leo and Cancer, the eyes the Earth produce see as well as the Earth can make them see. Likewise hands, which are ruled by Virgo, feet, which are ruled by Pisces, ankles, which are ruled by Aquarius, the heart, ruled by Leo, etc. In every creature, no matter where in the overall scheme of things, each and every part is a *miracle* of *perfection.* *Never* mere adaptations or make-dos. Because the Earth is made up of Leo and Aquarius and Cancer and Taurus and all the rest, it expertly combines the various parts to make the creatures which inhabit the planet. Precisely the same can be said for the Earth's plants, gems and minerals. Will everything on Earth perfectly express its inherent Zodiacal sign energies? Of course not, because the Earth itself is not perfect. There is a great deal of, well, dirt, and always will be. A painter leaves paint on his palette, a baker leaves flour behind on his counter, etc. No process is without waste.

Planets

While Astrology may be *of* the Earth, may be *in* the Earth, may be the very *life of the Earth*, by itself the Earth and its astrology are *static* and *unchanging*.

In reality the Earth is but one of a number of planets in orbit around the Sun and therefore must by definition be in continuous and ongoing relationship with each and every planetary body in the solar system. Mars, for example, may not have a direct impact upon me, but, as a planet, Mars is more than big enough, more than close enough, more than fast enough, to have an impact upon the Earth as a whole. Just as I would expect the Earth to influence Mars.

These interplanetary relationships are by means of *harmony* and *resonance*. Mars has its own harmonies and resonances, its own "style" and "personality" as it were. The ceaseless interplay of Earth and Mars will excite, in each planet, those elements and energies which are in sympathy or harmony. In other words, Mars will, in one fashion or another, "illuminate" or identify the energies in the Earth which are in resonance with it. In a larger sense, every part of the Earth will react, in one way or another, to the energies of Mars, taken as a whole. In other words, speaking broadly, Mars likes Aries and Scorpio the best, gets on well with Capricorn, but dislikes Libra, Taurus and Cancer.

As with Mars, so with all the planets. In addition to whole signs, there are various sub-frequencies, among them, day/night rulers, triplicity rulers, decans, faces, terms/bounds, dwads, etc. Indeed, we can say, collectively, that we only know Terrestrial energies, the signs of the Zodiac themselves, by means of their resonance with the various planets. Energies in the Earth which are not in some way "triggered" by the various planets will therefore be latent, unexpressed and unknown.

So we now have an Earth with its own unique Zodiacal energies, which are stimulated, excited, *brought into being*, by the various planets, the resulting vibrations so thoroughly mixed that it is pervasive in every rock, every spade of soil, in the very clay of our own bodies. While it is true that Mars has no *gravitational* effect upon any of us, it is also true that *because* we are made of the Earth, *because* we are made of material which is itself vibrating, we not only resonate to the Earth's own vibrations, but, we, each of us, also vibrate directly and individually to Mars. And, as before, with all the other planets.

Is distance a factor? Yes it is. Broadly speaking, the closer planets will be more nuanced and more detailed, with the Moon being the most nuanced and the most detailed. And in fact, in astrology, the inner planets (Sun, Moon, Mercury, Venus, Mars) are known as *personal*. Jupiter and Saturn, which are much further away, are *social*. The outer planets are *generational.* Outer planets only "step forward" when in close aspect with inner planets.

We have now arrived at a second fundamental principle:

The interplay of Earth and planets produce astrology as we know it.

The signs of the Zodiac are *in the Earth.* They are *brought to life* by the Earth's ceaseless interaction with the other planets. This makes the Earth a giant vibrating sphere, and it is on this sphere that we live.

A number of things have now become quite simple. Why is birth a critical moment? Here is an analogy. Imagine the Earth to be like a vibrating, orbital sander. Orbital sanders are hand-held electrical devices for smoothing wood. They have a flat pad on the bottom where the sandpaper is secured.

Take the sandpaper off and turn the unit upside down, that is, with the flat side up. Put a penny on the sander, and turn it on. The vibrating sander is the Earth. The penny is you. Both the sander and the penny are vibrating, with similar, but as we can see, not quite identical vibrations.

Now consider the Earth's vibration, unlike that of the sander's, is not consistent, but variable, and this because of the Earth's ceaseless relationship with the Sun, Moon and other planets. With the sander in your hand, carefully tilt it this way and that. Note how the penny reacts to every change. Move too abruptly and the penny will fly off the sander. Which, by analogy, is to say that when the Earth's vibration becomes too intense, we have accidents, we sicken and die.

Replace the penny with, say, a cookie and turn the sander on again. Run the sander until the cookie has become so distressed that it starts to break up. Note how each separating fragment takes on the sander's vibration as of the moment of its separation. Note how the main fragment changes its vibration after every loss. *This is a precise analogy to a mother giving birth to a child.* Note the subsequent interactions among the various pieces. Based on their location, mass and the specific vibrations as of the moment of separation, pieces will collide with other pieces at stated times and places. Astrologically this is known as synastry.

The initial vibration of all objects age, or *decay,* over time. In astrology, we "age" an individual's "vibration" (aka natal chart) by means of primary directions, secondary directions and solar arcs, among other means. (In India, by Vimshottari dashas, etc.) The relationship and interaction of decayed vibrations and transits is therefore obvious.

Note in this theory, a tree, precisely because it is firmly connected to the Earth, cannot be said to have a "birthday" as it never separates from the Earth.

Aspects

We might say that, yes, there are aspects between the planets. Why wouldn't there be aspects, when planets come to simple angular relationships among themselves? Regrettably, there are two problems with this.

One, while we have established the Zodiac to be in the Earth itself, this Zodiac is based not on longitude, but rather on the *Sun's declination.* If the Sun is at 21° 37' S declination, then its longitude is 22° 13' 57" of Capricorn and the date is more or less January 13 of any given year. (The leap year messes up the date slightly, this example is from 1950 and a midnight ephemeris.)

Moreover it is not entirely clear if this declination to longitude exchange should be based upon the Earth (geocentric) or the Sun (heliocentric).

The second problem is that Ptolemaic aspects, for heretofore mysterious and unknown reasons, are all centered around 60°. To these, George McCormack adds 30° and 45°, which is to say, half of 60, and half the distance between 30 and 60. With 30 we are immediately reminded of the twelve houses, which, ideally, are 30° each, which then reminds us of the signs, which, when projected into the sky, are also 30° each. We are being led, or, perhaps pushed, into some sort of definite structure. And it seems that we have a solution.

Researchers Ronald Cohen and Lars Stixrude, at the Carnegie Institution of Washington, have recently postulated the central core of the Earth to be a single giant crystal. Exactly what kind of crystal they have not yet determined, but they are leaning in favor of the crystal being hexagonal. Which is six-sided.

And in fact there is a variation of iron and nickel that forms hexagonal crystals. It is found in meteorites. Which, as everyone knows, are the remains of destroyed planets, presumably their cores. That substance is hexahedrite.

Upon learning of hexahedrite I was at first disappointed, as I wanted the Earth to have a twelve-sided crystal. One that would explain twelve signs and twelve houses, etc. But while looking at pictures of crystals online, I glanced at one too many pictures of crystalline water, in other words, snowflakes. And then it hit me.

The hexagonal crystal does not express six, but rather, twelve.

How does a six-sided crystal express twelve? Quite simply. Look at a crystal. You will find both facets and edges. Six-sided crystals have six of each. We might arbitrarily label them as six masculine edges and six feminine facets, or vice-versa. Which, to the Earth, are the six masculine signs (Aries, Gemini, Leo, Libra, Sagittarius, Aquarius) and six feminine (Taurus, Cancer, Virgo, Scorpio, Capricorn, Pisces). The hexagonal crystal at the center of the Earth, precisely because it is hexagonal, describes astrology as we know it. Describes the twelve-fold division of the Zodiac, describes the masculine-feminine polarities, and describes the innate nature of the traditional Ptolemaic aspects. All of which are based 30° and 60°.

This permits us to state a third principle:

Astrology may be defined as qualities inherent in, and the further study of, the six sided crystal at the center of the Earth.

Being the core of the Earth itself, the Earth's central hexagonal crystal is presumably aligned with the Earth's axial tilt of 23°44'. If so, then because this tilt is always in the same direction, we have the reason why 0° of Aries is set to the spring Equinox. The Equinox is the Earth's annual "re-centering" or "realigning" with the Sun. Projecting this hexagonal alignment into the sky lets us map the planets in the solar system from the Earth's own, unique, point of view. It gives us the geocentric, Tropical, Zodiac. We are mapping *the Earth's* relationship to the various other bodies in the solar system. We are not mapping the Sun, so we are not heliocentric.

Nor are we concerned with the Sidereal Zodiac, which is based, not on fixed stars, but on the Earth's axial wobble. This wobble is due to various gravitational influences on the Earth, most notably the fight between the Sun and Moon. Like as not, if there was no Moon, there would be no wobble and therefore no Sidereal Zodiac, or perhaps, one that was so slow in motion as to be impracticable. (The Sidereal Zodiac moves at the rate of 1 degree every 72 years, which is a cycle of 25,920 years, more or less.)

It is, in fact, the *projection* of the Tropical Zodiac, from the Earth, into the sky, that enables us to finally *identify* and *name* the Zodiacal energies *inside* the Earth itself. We do not know what the Earth's Aries energy may be, except that it is "illuminated" or "triggered" when planets pass through the section of sky which the Earth's Equinoctial settings have labeled as "Aries." This transference explains the traditional confusion about astrology, about what and where it really is and how it really works. It is almost impossible to see "Aries" in the Earth itself, but it is easy to see its *projection* into the sky.

So the *common* influence of the planets on the Earth, which is that Mars, for example, rules Aries and Scorpio, etc., becomes *dynamic* as the planet Mars moves about in space and appears to enter one sign after another. As it enters the various signs, as it shifts from "pushing" on a facet of the Earth's central crystal (at its spring alignment) to "pulling" on an edge between two facets, its vibratory impact upon the Earth changes accordingly. In common parlance, we say that Mars rules Aries from its current position in, for example, the sign of Leo.

Now note that the hexagonal crystal explains not only the twelve *signs,* but in the daily rotation of the Earth's axis, the twelve *houses* as well. Note that while the Zodiac is a matter of simple longitude, astrological houses are customarily calculated according to the Earth's horizon and Prime Vertical. *Not* (with the exception of Porphyry houses, an early and crude system) according to zodiacal longitude. With this minor tweak, we may now say that, for example, Mars in Leo in the third house rules Venus in Scorpio in the sixth, and *all of it* will be based on the *simple influences* of the planets Venus and Mars, *as they interact with the hexagonal crystal at the center of the Earth.* We no longer need starry constellations whatsoever.

It is when we take astrology *out* of the sky and put it *in* the Earth and *on* the Earth that we realize astrology's staggering power and its incredible capacity for detail. We no longer hesitate how astrology's mysterious, invisible influences can be transferred from one side of an empty sky to another, equally empty sky, seemingly on mere whim. We are instead witnesses to the inner dynamics of the Earth's own crystalline structure.

These on-going planets-to-crystal relationships are complemented by **special relationships,** known as *aspects,* which occur when the planets form harmonics between each other, as seen from the Earth and as framed and defined by the Earth's central crystal. When two planets form an exact aspect they create a vibration *external* to the Earth itself, but one which *strongly impacts the Earth,* by means of *vibratory sympathy with the Earth's central hexagonal crystal.* Blinded by the sky, intimidated by a phony science, modern astrologers have been guilty of overemphasizing aspects, at the expense of the Earth's more powerful, underlying astrological forces.

The various external planetary influences combine to change the vibratory rate of the Earth's central crystal, from moment to moment. The understanding and use of Astrology is obvious, logical, and necessary to all life on this planet.

George McCormack

We have now arrived at an understanding of McCormack's astonishing finding, that planets exactly on the midheaven/immum coeli (MC/IC, 10th/4th) have the strongest impact upon the Earth's weather. Consider that a planet placed on the IC will not only be passing its resonant energies directly *through* the Earth itself, but these energies, combined, enhanced and amplified by *precise alignment to the underlying crystalline energies of the Earth itself,* will emerge from the ground at *right angles to the surface.* In other words, *with maximum combined force.*

I remember an old rule from alchemy. The alchemical process, which is scoffed at by many, was said to require total darkness, or moonlight, in order to succeed. Moonlight is not merely very weak light, it is also *polarized.* Which is to say, light reflected over a great distance is polarized merely by the combination of *reflection* and *distance.*

If the Sun and planets (presumably also including the Earth's Moon) are crystalline structures, then the harmonic resonances that pass between them are presumably *polarized* as well. When these resonances arrive and then meet with the Earth in a precise polarity (the MC/IC axis), polarized-polarized results would be expected. McCormack terms this "magnetic action" (pg. 92 and elsewhere). I would refine this and call it *magnetic polarization.*

Planets are not magnetic to each other *per se.* They *become magnetic* when, in the sky, they are polarized at specific hexagonal angles. While holding those angles, they have specific impact upon the Earth's central crystal and its various discreet energies. Which that crystal then radiates in a general fashion to the planet as a whole, affecting all that live and breathe and crawl on its surface, and, specifically, radiates to key locations which affect the atmosphere and weather, if not local mundane events as well. Precisely as McCormack describes.

McCormack notes that planets that make 90° aspects to the MC/IC axis are themselves powerful. Which they should be. But note what he did *not* say: With the exception of the Moon, McCormack rarely bothers with the ascendant/descendant. Was this an oversight on his part?

No. Fairlawn, New Jersey, where McCormack made his observations, is 41° north of the Equator. At that latitude, the ascendant/descendant axis is rarely at right angles to the MC/IC. In terms of astrometeorology, McCormack did not find the Asc/Dsc angles to be of importance. Instead, he found the raw 90° angle itself to be supreme. This is a most important detail. McCormack is *not* describing astrology per se. He is, for the most part, describing the *astrology of the atmosphere.* The atmosphere is the part of the Earth with the least mass, it therefore has the least *resonance.* In practical terms, there may very well not be enough mass in the atmosphere for rulerships to apply.

In this regard I am reminded of the work of John H. Nelson (1903-1984). From the 1940's to his retirement in 1971, Nelson was employed by RCA in New York to identify and forecast times of radio interference. He quickly identified specific planet to planet *heliocentric* aspects as the culprits, eventually achieving a high degree of forecast accuracy. Why heliocentric? Consider that radio waves are broadcast from towers high above the Earth's surface, radiate into space itself and are presumably directly influenced by solar and planetary energies. Lacking all earthly matter, the Earth's own astrology would not seem to apply to radio transmission at all. It is very likely that Nelson and McCormack knew each other. I would very much like to know their opinion of each others' work, but I digress.

In studying McCormack's work, I gather that in the process of deducing that high and low pressure areas involve exchanges in the levels of the atmosphere, he was first drawn to highland and lowland areas as where these exchanges would be most clearly felt. He then presumably extended his observations to waterways and finally to sandy soil. In the process he stumbled upon something else of extraordinary interest.

It seems he accidentally hit upon a connection between western astrometeorology and Chinese Feng Shui. The term "Feng Shui" translates as "wind/water." It is the Chinese system of landscape management. It is also directly tied to Chinese astrology.

Chinese astrology is different from Western astrology, or, for that matter, Vedic astrology or Persian astrology, and this is because of number.

For many centuries astrology remained in the tropics, principally in India. Alexandria, at 32° N, was a lonely outpost. The reason was because astrology needs a house system in order to best express itself, and house systems are the simplest and work best in tropical latitudes, where houses and signs tend to overlap. Leave the tropics and the two systems diverge. By the time you reach northern Europe — or central China (Beijing is 40°N) — simple equal houses will not work, nor will trisections in zodiacal longitude (Porphyry) work. A proper house system trisects the Prime Vertical, either in space, or by time.

To do this handily, you must have a sophisticated number system, which the ancients lacked. Instead, they had hash-marks, the best of which seems to have been the Roman system (I, II, III, IV, etc.). Complex computations with hash-marks are not impossible, but they are unwieldy, which is why many ancient peoples employed the abacus to do sums.

Starting around 300 AD, India developed a number system based on place values, which we now call Arabic numbers. These were adopted by the Arabs in the 9th century, and in 1202 were introduced to Europe by Fibonacci. This single event revolutionized all Europe, but, regrettably China was left out. China did not adapt Arabic numbers until the 20th century.

To this day China has a variety of crude number systems, which is why it retained the abacus centuries after it had been abandoned elsewhere. China was too far north to equate houses with signs, so, lacking numbers, China was unable to erect astrological charts. Instead, they developed an entirely different system of astrology, one based on the 60 year cycles of Saturn and Jupiter. (From which, by the way, they get their five elements: Jupiter's cycle is 12 years, it makes five cycles in 60 years, while Saturn makes two.) By this reasoning, Feng Shui is the expression of the Earth's raw Zodiacal sign energies as they directly manifest in the environment. The combination of a crystalline Earth-based system of astrology, George McCormack's astrometeorology, and Feng Shui, may produce surprising results.

Astrology is under our feet. It radiates from the ground up. We are soaked in it, always have been, and always will be. We can no more deny it, than we can deny gravity. — *April 24, 2012*

◐ PEOPLE as TRANSITS

A couple of days ago I was asked for a book that would explain Parent/Child and Teacher/Student relationships. It's not in Davison's *Synastry*, or so the email said. Which is true, Davison has no separate sections. So I suggested Lois Sargent's *How to Handle Your Human Relations* and Lois Rodden's *Mercury Method of Chart Comparison*, but my querent already had both of them.

Sargent's book has chapters on Parent/Child, Family Relationships, Friendships, and Business and Professional, but the only chapter that's worked out in astrological detail is Business.

Rodden, on the other hand, says, *If you can't talk to them, what good are they?* Which is what you might expect from a Gemini with Aquarius rising.

But this was unfair to Davison. Look closer: He remarks that, for example, it's a bad idea if your doctor's Saturn falls in your 6th house: *Unless Saturn is well aspected at birth and the cross-aspects are particularly favorable, this is not a good position for an employee's Saturn, or for the Saturn of anyone who has care of the native's health.*

Davison makes these remarks where appropriate. Check your copy.

Which brings up the fact that people are living embodiments of the day of their birth, that they are *living transits*. What was July 6, 1946 like? Well, we should all know. We were once ruled by a man born on that day.

Davison's remark about a doctor's Saturn falling in my 6th is a personal sore point. My Saturn falls in my 5th, which makes me a true Don Juan (not!), which means the Saturns of doctors two years younger than me will always fall in my 6th. This is because Saturn spends about 30 months — 2½ years — in each sign, so, two years after I was born, that's where everyone's Saturn is, at least, so far as my chart is concerned.

Which in thinking about it, explains a lot about daily life. If, for example, your Saturn is in your first house, the Saturn of people two years younger than you will fall in your 2nd house. In your house of money, these people will be an economic drain. They will take all your money, one way or another.

Saturn in your first, the Saturns of people five years younger (two signs/two houses) will more or less fall in your third. What's the third house? Well, brothers and sisters, elementary education, short trips around town, the local environment in general. So don't expect them to get on with your brethren.

Your Saturn still in your first, the Saturn of that cute young thing, seven years your junior, will fall in your 4th house, of family and home. Don't bother introducing her to your folks, they will think she's beneath you.

The concept should by now be clear. Saturn is a touchy planet which moves at serene, regal pace. If you know the ages of the people around you, among them, your boss, your doctor, your dentist, your employees, your barber, your teachers, etc., you will know how you fare with them. An ephemeris will help, but you can make a rough guess off the top of your head and not be far wrong.

Their Saturn in your first? They won't like you. Nothing personal, you understand.

Their Saturn in your fifth? Forget romance. Don't let them invest your money.

Saturn in your 6th? As Davison says, bad if they're a doctor or employee. Or a chef, for that matter. Chances are, chefs whose Saturns fall in your 6th house will be found at restaurants you don't like, as you will find the food tasteless.

Their Saturn in your 7th? Possibility of a long-term relationship, but not of the cheery sort. More likely, someone you don't want in your life.

Their Saturn in your 8th? I've come to associate that with greed. Note the difference with their Saturn in your second. In the second, they're just expensive to have around. In the 8th, they're actively trying to separate you from your money.

Their Saturn in your 9th? You may find them religious and dogmatic. Or scientific and dogmatic. Isn't the 9th about overseas travel and all that? Well, yes, but their Saturn in your ninth won't take you anywhere, except possibly as their prisoner. Possibly a rather severe teacher or professor. We've all had at least one of those. What was his age? How much older?

Their Saturn in your 10th? This is one to look out for as regards employers. Hard and cruel for the most part. Hard and fair? Only if their Saturn is well-aspected and you're very lucky.

People as Transits

Their Saturn in your 11th? You may have no friends save for them. Their Saturn in your 12th? A jailor or a warden.

Since everyone you know is going to have their Saturn in one or another of your twelve houses, you have no choice in this. Like as not you can define your various relationships, overall, from this single factor.

Do the calculation. Saturn takes 2½ years per house. In five years it moves two houses, in 7½ years, three; in ten, four houses, in 12½, five, in 15 years, it's gone halfway 'round your chart and is now in the opposite house. Don't forget to do this both before and after your year of birth.

As with Saturn, so with Jupiter, only with happier outcomes. Jupiter moves about one sign/house per year. If your Jupiter is favorably configured in your natal chart and if your Saturn is not a complete pain, you may well enjoy the company of people your own age.

Looking for a Sugar Daddy? How many years will that be? Well, in my case, three, which gets Jupiter from my 11th house, into my second: People who bring me abundance. Money. Or would except in my case their Saturns all fall in my 6th, of health and work. The choice between Jupiter and Saturn for me was between money and being healthy. I chose healthy and, no, I wasn't aware of the astrological factors at the time. Things just worked out that way.

How about Jupiter in 5? Good for romance? Well, y-e-s, but if you have a water sign in your 5th, he/she may also bring babies, so watch out.

Pay attention to the other planets in your natal chart. As Other People's Jupiters and Saturns sneak around your chart like this, they will bump into your natal planets and create mayhem thereby. In my case, not only are doctors two years younger than me, with their Saturns falling in my 6th house a bad idea, they're also bad because their Saturns fall on top of my Mars, which happens to be in my 6th house. Saturn and Mars hate each other. I once had a very nice Jewish acupuncturist, two years younger than me. I liked him a lot, but eventually stopped seeing him as his Saturn instinctively wanted to kill my Mars.

If someone's Jupiter falls in the same house as your Uranus, this may give them the possibility of touching off your Uranus, with explosive consequences.

So as you eye the people in your life and advance their Saturns and Jupiters around your chart, pay attention to the signs/houses where your own planets are located. When their Saturn falls in the same house as, say, your Venus, you may have a hard time getting rid of them. They will like you just fine, but you may not like them. This is not a matter of aspect, but conflicting planets in the same house. Your natal planet sets the tenor for the house in which it resides, for better or worse. Someone's Saturn landing on top of it will be their guest.

The overall framework is set, not by the host planet itself, but by the nature of the house it's in. If your Venus is natally in your 7th house, when a friend's Saturn is in that house with it, you've got a sticky relationship, since both Venus and the 7th are about other people. Put them both in your third house and you've got a pest underfoot.

Change Venus for Mars, put it in the 3rd along with your friend's Saturn, and you have the possibility of traffic accidents. Mars is a two-year cycle. You will need to know at least their month and year of birth.

Finally, in this rump, everyday form of synastry, look in your chart for tight outer-to-inner aspects. In my chart I have Moon conjunct Pluto, for example. Which means that everyone my age has a Pluto that falls smack on my Moon. It made for grim days in 5th grade and I was ever so much happier when I could escape into the larger world with people with a greater range of ages.

In my case, I get on best with people five and seven years younger, or five or six years older. A bit of detective work and you will discover why your friends tend to gravitate towards births in certain years and not others.

Of course, if you have your friend's date, time and place of birth, you can set a full chart and do a proper synastry and learn all manner of things, but for most of the people, we only know their age. You can learn quite a bit with just that.

Some people think you have to go halfway 'round the world to get Jupiter in a money house. Try this instead: Find the years (12 years apart) when Jupiter transited your second house and then go shopping for money managers born in those years. But as always, watch out for their Saturns.

Shouldn't you be watching out for their Uranus, Neptune and Pluto, too? Yes, but as these move so very slowly, among your contemporaries they won't stray very far from their natal placements. On the other hand, if you're horsing around with grandmas or grandkids (or greats-, etc.) you already know they're a heap different from you.

PS: So far as Jupiter and Saturn are concerned, these are the rudiments of the Chinese animal/year system. When you combine the animal with the Chinese element (fire, earth, air, water, *metal*), the result is Jupiter/Saturn. — *May 1, 2012*

The true Pioneer of Astrology in America.
— *James H. Holden and Robert A. Hughes*

☉ Introducing Luke Broughton, *1828-1898*

LUKE DENNIS BROUGHTON was born in Leeds, England, on April 20, 1828, at 10 am. He presumably came from a preacher's family as his brothers were named Matthew, Mark and John. As a boy he was taught astrology by his father. It was a lifelong passion.

Around 1854 he emigrated to America and settled in Philadelphia, where he joined his brothers Matthew and Mark, all plying the family trade. Luke graduated from the Eclectic School of Medicine in Cincinnati, as an herbalist and thereafter practiced astrological medicine. The combination of herbalism and astrology is profoundly powerful, but without the weight of astrology, herbalism will always be shouted down by the drug labs, peddling their awful poisons.

The last year of his life Broughton published *The Elements of Astrology*, which I will have in print in about two weeks. It was the first great astrology book printed in the US. It might be subtitled, *The Life of an Early American Astrologer*, as in it Broughton documents the raw hostility he faced in his practice and promotion of astrology: Mail opened, lecture hall destroyed, home ransacked. He was driven out of Philly in 1863 and settled in New York, where he did not fare much better.

This 500 page book reads a lot like one of my weekly articles: Direct, personable, immediate, with its flaws fully acknowledged and on display. This is the distillation of a working astrologer of a century ago, before there was astropsychology, Sabian Symbols, New Age, and all the rest.

In this book you will find the charts of Abraham Lincoln, U.S. Grant, Queen Victoria, William McKinley, William Jennings Bryan and many others who are still well-known. This is an enlarged facsimile edition, as big as I can make it. Hidden inside are many fascinating details.

On the next page, Luke Broughton's natal chart. Note the Moon-Saturn conjunction in the 12, Venus in the same sign as the cusp of the 12th. This man keeps his feelings, his sorrows, and his love, private. Note Mars opposite to Moon/Saturn in the 6th: He will be a workaholic and stir up resentment with the people he works with. Which will externalize and become personal because of the Mars-Moon/Saturn opposition. He can't help it. The act of living (Mars) generates hostility.

Mercury in 10 in Aries is a public speaker with a great deal of energy. — *May 1, 2012*

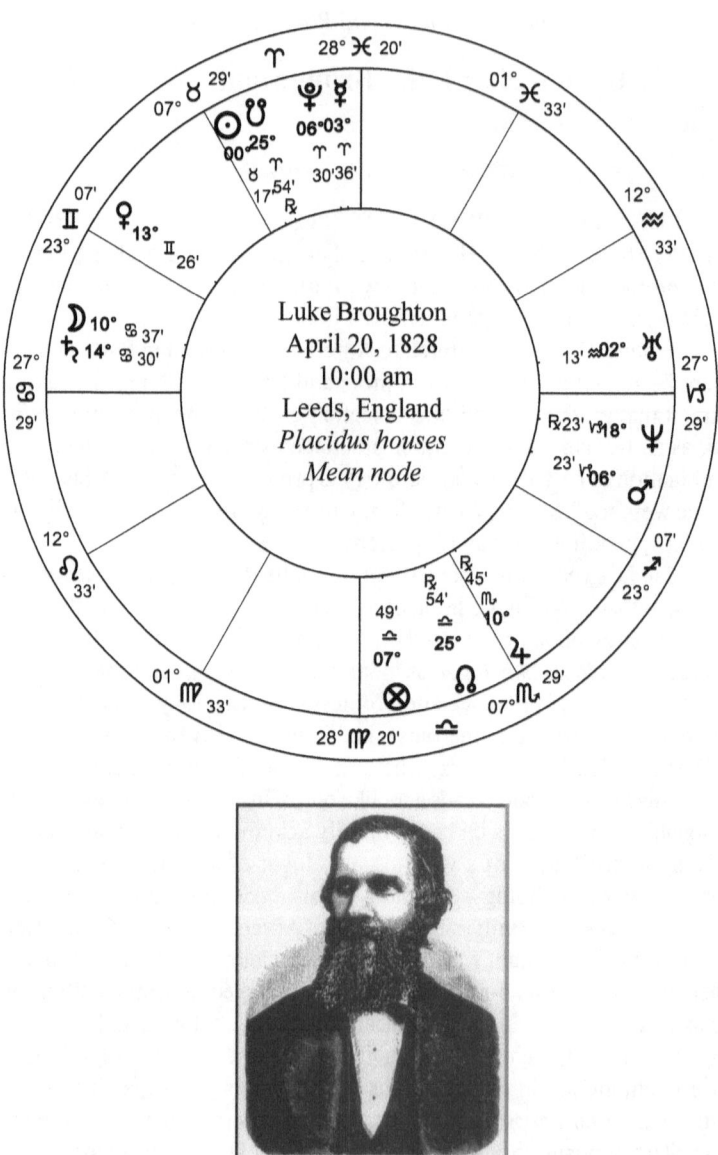

Everybody's favorite host:
☉ Tom Bergeron

I first saw Tom Bergeron as host of the revived Hollywood Squares, where he made the show sing, from 1998 to 2004. It ain't easy keeping nine maniac star egos stroked (especially when they're all over the hill and desperate for work), and make the contestants feel comfortable and make his own job look easy. I was surprised when the show was cancelled, but I've been around show biz long enough to know that ratings don't have a lot to do with these things.

Tom was born May 6, 1955, at 10:30 am, in Haverhill, Massachusetts. AstroDataBank says someone quoted the birth certificate, so it gets a Rodden AA rating.

Bergeron is a Taurus Sun, Scorpio Moon, Cancer rising with a debilitated Venus on the MC. Venus on the MC gives us our first clue: In a female nativity, an "exposed" (MC) Venus in Aries makes for a tomboy, someone proud of her masculine forcefulness.

It's the reverse with a man. Venus in Aries on the MC makes Tom timid. Uncertain of how he should express himself, hesitant that he not publicly "go overboard." But whenever Venus lands on an angle, we, as observers, find the affairs controlled by that angle to be harmonious. Which is to say, delightful. In other words, instead of pushing himself on us as The Host, a debilitated Venus on his MC makes Tom more than happy to let others shine. Which is the essence of a good host.

Next, we see that Tom is a full moon birth. I've already shown you many full moon births. You should know by now these are very dynamic people, able to cover a great deal of ground, capable of inventing and reinventing themselves. Full moons are often mistaken as bipolar or two-faced, and can transform everything they touch.

The house polarity is 11 (Sun) to 5 (Moon), which is to say, friends vs: romance, or, more broadly, making love to a group, vs: making love to one single person. Since people center, intellectually, around their Sun, and emotionally around their Moon, we can say that Tom, with his Taurus

Sun, is practical and levelheaded and likes his rich friends.

We can also say that his personal, lunar life, is frustrating. Moon in Scorpio has intense desires, but, applying to Saturn, those intense desires are always frustrated. Intensely so. Moon-Saturn conjunct in the 5th, Tom's ideas of play are stilted. Not surprisingly, the details of his private life are, well, private.

As the 5th house is also the house of fun, and as people tend to stay longer in jobs they like than those they don't, we might note Tom's longest running hosting job has been with America's Funniest Home Videos, 2001-present. When Tom hosted Hollywood Squares, the selection of stars and contestants was largely out of his control, so he had to make do as best he could.

That's not the case with America's Funniest. The host would be expected to have final say on what airs, and what does not, as in, "Tom, do you think you can punch this?" ("Punch", as in *give it a good intro.)* Result: What Tom likes goes on the air, what he doesn't like, doesn't. After a season or two the producers will know his likes and dislikes and cater accordingly.

Watch America's Funniest and you can learn a lot about Tom, in particular, his Moon-Saturn Scorpio conjunction on his 5th. Here's one of Tom's own creations: "*'Head, Gut or Groin', where Tom picks two members of the studio audience to guess whether the person in the video will be hit in the aforementioned three areas of the body (though occasionally, a video in this segment may feature a person getting hit in two of the three areas)* ..." (From the Wiki entry on America's Funniest Home Videos.) Remember these are viewer submissions and we may presume there are many submissions too horrific to screen. (Remember MTV's Jackass?)

By contrast, here is what AFHV did with its first host, Bob Saget (1989-97): *"Three $100,000 contests air each season. . . During the Saget era, the set would be decorated with balloons; and beginning in the second season, a revolving gag involved the "money" being guarded in some bizarre way from Saget on stage, including a security guard or a force field. Once the winner is announced, a marching band would often appear on stage playing the theme song (other times the regular theme would be heard), and balloons are dropped from the ceiling. All of this was scrapped after Saget left the show."* (Ibid.) Saget was born May 17, 1956, Philadelphia, time unknown. Pity. I would contrast his 5th house to Tom's, if I could. (Saget grew bored with the show, performed poorly, and left when his contract expired.)

With dispositors we can put things in motion, which leads to surprises. Sun in Taurus is disposed by Venus in Aries. The two are in mutual reception. This smooshes houses 10 and 11 together, which is to say, profession and friends become as one.

Yet, Venus is properly disposed by Mars, in Gemini. Now note that Mars also disposes Moon and Saturn, both in Scorpio.

So in Tom's chart, the Sun is disposed by Venus, while the Moon, which also happens to be the chart ruler, is disposed by Mars, and the two dispositors, Mars and Venus, are themselves both elevated, and in sextile to each other. Additionally, to Moon/Saturn, Venus and Mars form a Yod.

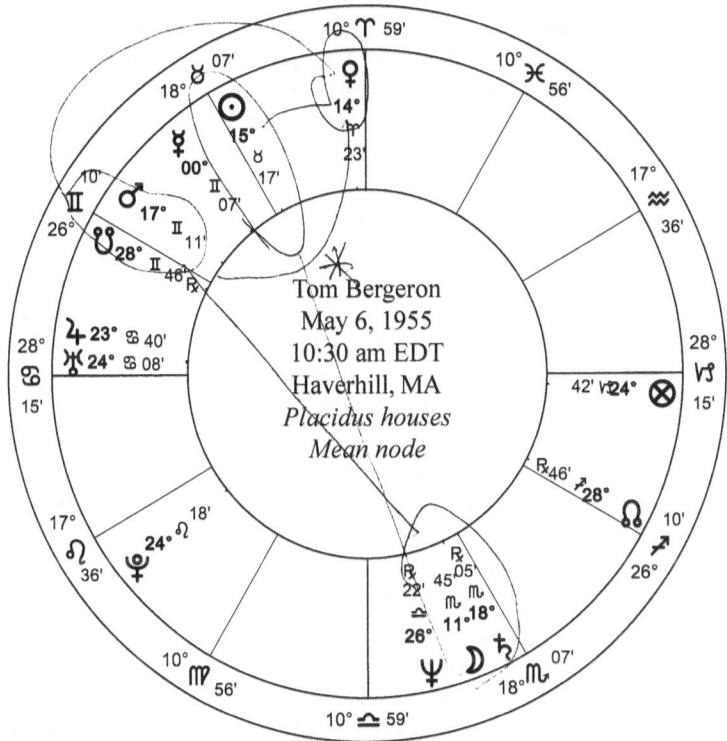

Tom Bergeron
May 6, 1955
10:30 am EDT
Haverhill, MA
Placidus houses
Mean node

Astrology texts say it's good when an opposition has a resolution like this, and Tom has it both ways. Once as an aspect pattern, albeit a rather strange one, and, second and more important, as dispositors. Dispositors can be thought of as planets that *lean on one another.* Give themselves mutual support. This is particularly important when the opposition in question is the full moon, as it is the King (Sun) and Queen (Moon) not only of oppositions, but of all aspects.

I myself have some of that with my full moon, and while, yes, I find that things do "resolve," that resolution is largely *exterior*. The person himself is still conflicted, is still forever of two minds, this way or that.

In Tom's case, this conflict can be seen, quite graphically, not in his career successes, but in his failures. Locally, Bergeron was an early success in radio and eventually got picked up, first by local TV, and then by the majors.

His first nationally syndicated TV show, *Breakfast Time,* on FX, was not shown in his home town of Haverhill, and this despite strenuous local efforts. When the show moved to Fox, it was revamped and Bergeron dropped.

Tom then moved to ABC, where he was a guest host on Good Morning America, but failed to get a permanent gig. And he is, or was, a "fill-in" host for the US version of Who Wants to be a Millionaire.

Instead, Tom got the syndicated Hollywood Squares, the syndicated AFHV, and the ABC show, Dancing with the Stars, where his role as host is primarily as a stabilizing and connecting thread. Tom has just turned 57. In terms of career, he is at his very peak, but he has consistently failed to get the big prize. Why? My guess is that his full moon, which should give him the power, is weak.

The Sun has a weak ruler, while the Moon is in an unfriendly sign and is applying to Saturn. Their ruler is sailing into the obscure 12th house. Like most 50-somethings, Tom has done his best, or as he himself put it in the title of his autobiography, *I'm Hosting As Fast As I Can!*

Second-run, second string, so close, a heartbreak. Will he ever break into the big time as the next Merv Griffith or Oprah Winfrey?

Look at May next year, 2013. Uranus will conjunct his midheaven (presuming 10:30 am is more or less right). Jupiter will conjunct his Mars. Saturn will be backing away from a station on his Moon, which Pluto will be sextiling. The solar eclipse of May 10, 2013, at 19 Taurus, will miss his Sun/Moon, but will tightly oppose his Saturn. Tom will clearly have an interesting time next spring, but probably not a happy one. Which means I, for one, think a major career advancement at that time (Uranus MC) to be unlikely. Tom Bergeron is a sweet, kind, likeable man, but Regis Philbin he is not.

Tom has cardinal signs on his angular houses. This normally indicates a powerful individual. Let's look closer:

Cardinal Cancer rises. It is ruled by the Moon, but the Moon is debilitated in Scorpio, moving into a conjunction with Saturn, which the Moon personally dislikes, and is opposed to the Sun. It's had happier days.

The descendant is Capricorn, which is an austere, "older" partner. Ruler Saturn is in Scorpio, where it is harassed by the Moon. In Tom's 5th house, partners will not be fun nor romantic. In his partner's 11th house (turn the chart), Tom's wife, *so far as Tom is concerned,* has few if any friends.

The MC/IC axis is not much cheerier. Aries is on the MC, ruled by Mars, which we find in Gemini. Mars is not badly placed in Gemini, but there are better locations. Mars rules iron (a physical substance), he likes things simple and clear cut, as he himself will eagerly do the cutting. Gemini is of the air, which strands him, and is dual, giving Mars a choice that it does not want. By house, Mars is in the worst of all possible locations, the 12th, where he cannot be seen at all. (This in keeping with my view that planets in the same sign as the cusp, want to be in that house.) And, finally, the MC is cluttered up with Venus. Venus and Mars do not like each other. Scratch the MC.

The weakest of the four angles (not just in Tom's chart, but in astrology in general) is the IC, or 4th house cusp. There we find Libra, ruled by Venus, which we find, debilitated, in the 10th. Planets in debility want to be in the house opposite. In Tom's chart Venus will act as if it's in the 4th, but this rarely amounts to anything as debilitated planets have not the strength. Tom presumably wants an elegant home, one that his wife (Libra = partner) can be proud of, but I would be surprised if this has ever quite come to pass. Note also that Neptune in Libra clutters things up. Which is another way of looking at the fogs of Neptune: As clutter.

Well then, what about Tom's full moon? As we have seen before, full moon are intensely powerful in their own right. Let's look more at Tom's:

Sun in Taurus is disposed by debilitated Venus in Aries. So when the Sun goes looking for support, Venus won't give him much, and, as Venus is in the most exposed section of the chart, its inability will be on public display.

The Moon in Scorpio is debilitated by sign to start with. It is further hampered by Saturn, as well as by the inconjunction to its ruler, Mars. In my view, inconjunctions are aspects of invisibility. When the Moon, which is debilitated, needs support (which it does), it will not know where to turn to find it, because its ruler is hiding. Hiding because of the inconjunction between the two, hidden again because of its house placement, the twelfth. A 12th house Mars, Tom plots in secret.

You will note the Moon is trine to Jupiter and Uranus, in Cancer, the Moon's own sign. Will this help? Increasingly I think planets in trine are like good friends who cheer you on. You still have to do the work. And as I've peeked at Wiki's Millionaire page, trines, to me, sound like the "phone a friend" lifeline. Good for moral support, not so good for real help and advice.

My best guess is that next year Tom Bergeron, a nice, likeable guy, will suffer a career reversal. I hope he can recover.

In this survey of Bergeron's chart, note that while Jupiter is exalted in Cancer and very near the ascendant, its ability to make Tom "larger than life" are unfortunately cut off by Uranus, which stands in the way. Also note his gift for gab is from Mercury in Gemini. Which, through its rulership of Mars (in Gemini) gives it the additional rulerships of Venus, Sun, Moon, Saturn, Neptune, etc. Tom's Mercury has no aspects per se, but its rulerships cover a lot of ground. This gives Tom great wit and an easy manner.

You are wondering if I will conclude by giving you Tom's expected life. No. Spectators at the Coliseum came to see strangers clawed and hacked to death. They did not care who the victims were, only that they were amused for the afternoon. I must be careful not to turn this into a Coliseum. Despite his best efforts on America's Funniest, Tom is an innocent. In reality, aside from idiot doctors, there are precisely two sorts of people who pronounce in this fashion: Astrologers and judges. The pronouncement is not of life, but of *death.* Aside from some wild west jurisdictions with trials in the mornings and hangings in the afternoon, such pronouncements are conditional. But they are, in the end, *final,* as well as horrible.

Tom Bergeron will leave us when God calls him home. I wish him a long and happy life. — *May 8, 2012*

☉ Venus, Money and the Eclipse

Venus goes retrograde on Tuesday morning at 10:33 am EDT, just shy of 24 Gemini. Venus rules money and one way you will always remember this is when, in March, 2009, the stock market bottomed.

Stocks had fallen uncontrollably since October the previous year. Why did the decline halt on March 9, 2009? Economic wizardry? Hardly. On Thursday, March 6, Venus went retrograde at 15 Aries. The stock slide ended the following Monday.

Isn't that strange? A retrograde Venus causing stocks to *stop* falling? Would not a retrograde Venus do the opposite? Make a falling market fall all the faster? Is it not true that planets in direct motion support good things, and that when they go retrograde, that support stops and weak or unstable things might unexpectedly collapse?

I would agree to all of that, but here's the catch: You're thinking that Venus direct was supporting *honest* money. The abrupt market reversal in March, 2009, hints that honest money had all been lost and that *phony money* had replaced it. So the retrograde was like fertilizing noxious weeds in the garden. They grew like crazy. Of course the market went up. It was growing weeds.

Sunday/Monday's solar eclipse (May 20, 7:47 pm EDT) starts in Texas, sweeps northwest to the Aleutians before heading back down the Japanese coast and ending over Hong Kong. It is at 0 Gemini. It will be the first eclipse to conjunct the Pleiades since they entered Gemini a few years ago. Robson says eclipses intensify the innate nature of fixed stars. The Pleiades are malefic, the eclipse is an omen. So far as I can tell the eclipse will not trigger another quake, but whatever it brings, it will not be good. It looks like a long week. — *May 15, 2012*

July 25, 2012: Typhoon Vicente struck Hong Kong at 4 am on Tuesday, July 24, 2012.

☉ The New Medicine

Modern medicine we all know quite well. We have an ailment, we go to a doctor, we describe our symptoms. For the most part, the doctor has no idea what the problem is. Many diseases, with quite different causes and very different cures, have similar symptoms.

So he takes our temperature and our pulse and runs a battery of expensive tests and then after consulting his standard reference (*Physician's Desk Reference*, or PDR) makes his best guess and prescribes some sort of pill (or sometimes liquid) that is supposed to make us better.

This guesswork can result even when the ailment is quite highly localized and very acute, as we learned a year ago with my wife's sudden and severe cornea inflammation, which, more than one year later, remains officially undiagnosed. I used both the natal, as well as the decumbiture chart (chart for the time the patient takes to his bed, or, anymore, when the ambulance arrives) and diagnosed the problem in minutes. It was the left eye, which made the Sun the agent as the Sun rules the left eye in females. In this case, the transiting Sun exactly squared Pluto, which was making a station directly over my wife's Saturn, both in Capricorn. With Sun/Aries and Pluto/Saturn/Capricorn, there was more than enough for a competent herbalist to work with. Regrettably, Johns Hopkins are not herbalists, they are not astrologers, and they are still clueless.

While I was distressed and shocked at the poverty of modern medicine (Hopkins is world-famous), the only surprise, to me, was that a lunar-based decumbiture chart did not work. A solar decumbiture did. Crisis occurred every three months, like clockwork, when the Sun squared itself. Solar /lunar decumbiture, what's the difference? Only which light you are following. It's otherwise exactly the same chart.

On to this week's topic:

The New Medicine

The new medicine is the old, old medicine: Herbalism of one sort or another, which can be naturopathy or homeopathy or whatever. Mixed

The New Medicine 175

with a dose of what used to be called *"Eclectic,"* which is to say, whatever works. Joseph Blagrave was an eclectic practitioner, for example.

Why was herbalism pushed aside in favor of patent medicines, and pills and elixirs? Two reasons:

First, when we're sick, money is no object. We want the best and we will spend as much as it takes to get well again. Any herbal doctor can grow half the herbs he needs in his own private plot, and easily get the rest from speciality houses. Herbs are cheap. Always have been. In the inverted world of medical pricing, there's no way cheap herbs can compete against expensive, factory-made poisons.

The second reason herbalism failed — and always will fail — is because herbalism does not have astrology behind it. Without astrology as a foundation, a herbal doctor is just another bungler. Just another guesser. He has no more idea what the problem is, than a fancy AMA-board certified doctor. Herbal cures are only superior if the diagnosis is correct. A good herbalist will get the diagnosis correct eventually, but that's also true of a well-experienced doctor.

In medicine, you don't want "eventually." You want, **"Yes, ma'am. I've seen this before, I know what it is, I know what to do, I have a cure, I will make you whole again."** *Without the least hesitation.* And that's what astrology gives us. *Astrology is the essential organizing principle.* With it, patterns are established and the world comes into focus. With astrology, we have a hierarchy, we can establish our position relative to others around us, we can work with knowledge and confidence.

Without astrology, everything is scattered and meaningless and all we can do is guess. Which puts us at the mercy of those with big mouths and bigger egos. Once real knowledge is lost — as it has been — the world becomes utter chaos. It's not that I have scorn for Gregory House, and his foolish, dangerous guesses, but that I have contempt for the society in which he lives and works, a society which has made this man a hero.

Astrological doctors have long claimed that astrology pinpoints the cause. Among those making this claim are Joseph Blagrave, H.L. Cornell, Richard Saunders, Nicholas Culpeper, Luke Broughton and William Lilly. Why is this?

Prior to me and my theory of astrology (the *Earth-Centered Theory*, I suppose), it was simply presumed that *As above, so below,* that there was some sort of murky relationship between stars and planets in the sky and the ailments we suffer from down here.

In placing the signs of the Zodiac in the Earth itself, in the very clay of the Earth, in making them Earth Energies, in assigning to the planets the role of triggering these Earth Energies in various ways, I have transformed the entire subject. I have transformed astrology into an exact and potent *Earth science.*

Our bodies are living mixtures of astrological clays, clays that, by definition, contain and express zodiacal energies, each of which has its own particular quirks, its own particular strengths and weaknesses, and all of which are subject to ceaseless planetary influences.

Disease is then nothing more than astrological energies breaking down in astrological ways. Mars producing fevers, Venus inducing gluttony, Saturn old age and debility, etc. All in foreseeable, predictable, *mechanical* fashion.

The *natal* and *decumbiture* charts tell us the exact nature of the problem, for problems can only arise *if* and *when* astrological factors show them.

By contrast, medicine believes in a bacteria/fungus/virus theory of disease. Treatment cannot begin until one of these three has been identified as the culprit. But these agents cannot strike unless the body is weak, and it is astrology which shows the weakness. Moreover, as we learned with my wife's eye inflammation, the bacteria/fungal/virus invader theory is often useless, as many ailments, including my wife's eye inflammation, have *no such cause.* Johns Hopkins failed to diagnose my wife's problem. I did not. Just as there is nothing that cannot be expressed with language, just as there is nothing that cannot be shown in the natal chart, there is no disease, no ailment whatever that is not a product of astrological malfunction. Since astrology makes up the Earth itself, there are no exceptions, there can be no exceptions. Astrology supersedes theory.

Since these facts are true, then just as diseases are discovered by astrology, their cure, *the cure for all diseases,* is to straighten out the underlying astrological problem.

Given that the earth itself contains highly concentrated versions of pure planetary energies (metals, crystals, herbs, woods, clays, etc.), and that these expressly control the Earth's own zodiacal sign energies, then the skillful application of these materials will cure the problem, presuming the ailment is not in fact fatal. (Which as I have shown, astrology can also determine, though the practitioner may well shy away from burdening the patient thereby.)

When the subject is put this plainly, it is clearly stupid *not* to use astrology. Yes, your university trained doctor will sneer at you — and me — all the while shaking his head and throwing up his hands. He *firmly believes* that has been taught all that is known. If *he* does not know, then it *cannot be known.* Nor will successful astrological demonstrations change his mind. He is blind. You will meet many like him.

But you do not care. You are ill, you want to be cured, you have a right to competent medical care, you do not care if the doctor is a scientific wizard or a Siberian shaman. So long as the terrible disease ends and health and vitality returns. That's all you want, and you should never waver in your determination to get exactly that. Which leads to:—

The Training of the New Doctor

Medical training starts with Astrology, and by that I do not mean astro-psychology.

For natal astrology, **MORIN**.

For horary, William **LILLY**.

No, it does not need to be those precise authors. Patti Tobin Brittain has a brilliant take on Morin. John Frawley, Anthony Louis, Ivy Goldstein-Jacobson, Derek Appleby and many others all have brilliant books on horary. Find them in the bibliography.

And it's not just that doctors are to be trained with astrology, but that Astrology, being *the study of the energies inherent in the Earth itself,* will be the required foundation for all sciences and all disciplines.

Next comes preliminary medical studies, which includes H.L. Cornell, Joseph Blagrave, Richard Saunders, Nicholas Culpeper, Lilly (again), as well as the superb 19th century classic, *The Smith's Family Physician,* of 1873.

You will say that disease is not the same now, as it used to be. We have many more categories, many more diseases. This is an illusion. What we have are many competing egos, each eager to attach their name to some esoteric variant and attain personal immortality thereby, combined with a medical industry eager to make big bucks by micro-managing diseases with specific, and expensive, tests, machines and patent cures.

What we have, in reality, are the same diseases we have always had, based on the same dysfunctions of twelve zodiacal signs and seven *(plus)* planets. It is therefore easier to learn, or relearn, the traditional names and terms. *Smith's* is excellent for this, and for many other mundane purposes.

Learn the original, common terms and all the great medieval texts, up to and including Cornell (1932) will open up in front of you.

Only at this point will the candidate take up formal medical training. At this point it makes no difference what school of medicine he chooses, though I rather suspect the modern, state approved schools will not interest him.

The New Treatment

The doctor sets a chart upon the arrival of the patient. This is his decumbiture. Standard medical intake forms request the birth date, which to modern doctors is a trivial detail. The astrologically-based doctor will ask for place, and time of birth, if known. If there is no birth time, an experienced doctor will rectify the chart based on appearance, as suggested by Blagrave. Doctors deal with physical bodies, they are well-able to judge physical appearance in astrological terms, if they are only so trained.

Having the patient, his symptoms, his natal chart as well as the decumbiture, the new doctor will know immediately what the problem truly is, and will know the precise astrologically-based remedies to apply.

Presuming he is using herbs of one sort or another (which he will in many cases), those herbs he will have personally harvested, per Blagrave, during the proper planetary hour. They will be "super herbs."

After the initial treatment, follow-ups will be on days determined by the decumbiture, and, per Broughton, in a beneficial planetary hour. Doctors often complain of long hours and unhappy patients, but it is their own fault. Astrological diagnosis and cure results in certainty, while the decumbiture allows for organization. If it's Tuesday, I need to check Mrs. Bowlinger, as today's her day. In anticipating a stressful day, she will think I am a miracle worker, when all I am doing is following a simple plan. Instead of hodge-podge and ongoing emergencies, astrology gives organization and certainty. Order, not chaos.

Such is the New Medicine. Like as not there are a dozen more priceless medieval medical manuals waiting to be found, among them, the 16th century Dr. Foreman, who I suspect was the author of Richard Saunders' book.

Is this "new medicine" merely a repeat of the old?

I suspect not. I suspect it's never quite been all put together in one place and at one time. For example, both Lilly and Blagrave delineate Moon-Mars and Moon-Saturn aspects (also Moon-Mercury and Moon-Sun) in the decumbiture. Of the various possibilities, Lilly declares five or six to be fatal. With better skills and better herbs, Blagrave declares none to be fatal.

So, will this happen? Will there be a new medicine? I do not know. Western society may well collapse into penury and feudalism, and within the lifetimes of those now alive. Mere knowledge cannot prevent the mass blunders of greedy and stupid men.

Having recently published McCormack's weather book to great acclaim, it seems to me that a great many people are already ignoring science and using astrological weather forecasting to set the dates of upcoming events. The monolithic scientific "purity" which we presume exists, might well only be a facade.

One straw in the wind is NASCAR, which annually schedules nearly 40 races. This year, for the first time in 54 years, the Daytona Beach race was postponed for a day, from Sunday, February 26, to Monday, February

27, 2012. Checking the local climate, I see that Daytona Beach averages 2.77 inches of rain in February, and that in a typical February in Daytona a total of 7 days, 7 hours and 12 minutes experience measurable precipitation. Races cannot be run on wet track, the results can easily be fatal. Odds of rain on race day in any given year: Slightly greater than one in four. Actual postponements because of rain: Once in 54 years.

The conclusion is clear: NASCAR employs an astrometeorologist. Subtract 54 from 2012 and what do we get? Why, one of McCormack's original students, I'll bet! In 1958, McCormack had been a sensation for eleven years. NASCAR itself was founded in 1947, the same year that George McCormack became famous for his work.

By contrast, the scientific NASA, 65 miles south of Daytona, is often forced to reschedule launches, often repeatedly, due to weather. Are rockets more sensitive than race cars? Maybe, but rockets need only a 10 minute launch window, and with over half a century of experience, NASA's gotten good at squeezing launches into narrow time frames. NASCAR needs a solid four, preferably six hour chunk of dry time.

So when do we start?

When do you want to start? The need is great. The time is now, the place is here, the candidate is you. Suffering ends when you care enough to do your part, and no matter who you are, no matter where you are, there is a part that you can play. — *May 15, 2012*

🕘 The Development of Science

The story so far:—
Science, as we know it, traces back to the Age of Enlightenment, c. 1650, which came hard on the heels of the Peace of Westphalia (1648) which put an end to the 30 Years War (1618-48) which largely depopulated Germany and central Europe.

The Enlightenment was a product of the Dutch, French and English, who were unaffected by the horrible war. In formulating modern science, the Enlightenment declared urban beliefs to be "scientific" while denouncing rural beliefs as "superstition." You will find this division to still exist.

This division, urban/educated versus rural/ignorant, was particularly French and persists to this day, in the Academie and its well-known intolerance of the French popular masses, in its insistence, for example, for the academic *fin de la semaine* over the popular expression, *le weekend.*

Well, okay, we understand all that, we understand that French science is a city thing and the French are effete cheese-eating dandies and all that, but here's the catch:

The Enlightenment threw away the underlying Greek cosmology which had been carefully and lovingly revived over the previous 500 years, by the Italians, the Spanish, and the Germans, among others. As if Aristotle, Plato, Socrates and all the other Greeks had simply never existed. *How was that kind of blunder possible?*

Well, one reason would be that by 1650, Germany had been hacked to pieces, and so whatever they knew was at least temporarily unavailable. Which would let the French redefine the world any way they wanted, but that still does not explain this French blind spot.

Let's go back further. Let's go back to the Renaissance.

The Renaissance, in Italy, was an outgrowth of trade with Constantinople, coupled with loot extracted from the Holy Lands by the Crusades, which led to the rediscovery of the Greeks.

The Development of Science 181

Spain got to the Greeks by a different route. They were overrun by the Moors, who were Islamic. As was true then, as well as now, Moslems are great scholars, having based much of their culture on a careful study of the Greeks. Starting in the 12th century, a group of Spaniards set about translating every Arabic text they could find into Latin. (Known as the 12th Century Translators.)

And though, for the most part, it was a straightforward Arabic to Latin exercise, many of the Arabic texts they translated were originally Greek. Which set off a revival of Greek culture and philosophy. The spread of these texts to Italy (who were great traders) eventually touched off the Italian Renaissance.

The spread of the Italian Renaissance is of great interest. Wiki has a brief entry on the German Renaissance. Here are some extracts: *"The German Renaissance pushed **classical thinking**, arts, and the natural sciences to the forefront during this period of thinking with Germany. This also made scientists focus more energy on the world around them and focus less on the heavens."*

and, *"The greatest mark of the Renaissance was the renewed interest in **classical learning**. Documents, papal or not, were being brought to the surface for examination and study. **Classical learning** and study was a must for any person living in the renaissance and was considered a great part of one's education. The basis of literature and art in this time were references back to times with **Ancient Greek** and Roman societies and mythology. The basis of natural science developed from that same look back into **Greek** and Roman philosophies and teaching, however they were more developed."* (emphasis: Dave)

By **Greek** and Classical, what is meant was the wholesale revival of Aristotle by the Germans.

Now look at the Renaissance from the French point of view. Again, Wiki:

*Notable developments during the French Renaissance include the beginning of the **absolutism** in France, the spread of **humanism**; early exploration of the "New World" (as by Giovanni da Verrazzano and Jacques Cartier); the importing (from Italy, Burgundy and elsewhere) and development of **new techniques and artistic forms** in the fields of printing, architecture, painting, sculpture, music, the sciences and **vernacular literature;** and the elaboration of **new codes of sociability, etiquette and discourse**.* (emphasis: Dave)

Nothing classical here at all!

My highlights, above, roughly duplicate Wiki's links. *Absolutism* was the absolute authority of the French King, which would reach its apex in Louis XIV and which the Academie copied wholesale. Humanism comes next. That's scientific, is it not? Let's have a look at it:

Wiki says that Humanism started in Italy, and as Humanism gets a mention under French Renaissance, that's where the French got it. What was Italian — and then later, French — Humanism? Wiki:

Once the language was mastered grammatically it could be used to attain the second stage, eloquence or rhetoric. **This art of persuasion** [*Cicero had held*] *was not art for its own sake, but the acquisition of* **the capacity to persuade others** *— all men and women —* **to lead the good life.** *As Petrarch put it, 'it is better to will the good than to know the truth.' Rhetoric thus led to and embraced philosophy. Leonardo Bruni (c. 1369-1444), the outstanding scholar of the new generation, insisted that it was Petrarch who 'opened the way for us to show how to acquire learning,' but it was in Bruni's time that the word* umanista *first came into use, and* **its subjects of study were listed as five: grammar, rhetoric, poetry, moral philosophy, and history."**

Contrast to the Germans and the conclusions are shocking.

The *Germans* saw the Renaissance as a revival of **Greek cosmology**.

The *French* saw the same movement as *artistic,* based on a revival of the **Latin language.** They took this language, combined it with Absolutism, to create "the capacity to persuade others to lead the good life." *Et voilà. Le fin de la semaine,* like it or not. The subsequent Enlightenment was the French dictat to the world. Or to be still more precise, one of a long series of dictats, by a society that still believes in absolutes.

German vs: French. Greek vs: Roman.

The Germans came to Italy to *learn.*

The French came to conquer. Which, in fact, they did. Knowing the French a little bit as I do, they believed themselves to be culturally superior beings. (Rather like modern Americans.) The French did not come to Italy to learn. They came to *copy* and *extract.* François Premier's encounter with Leonardo was typical. Further in Wiki's article on Humanism is this: *Renaissance humanists, who considered themselves as restoring the glory and nobility of antiquity, had no interest in scientific innovation.*

It is now clear what happened during the 30 Years War: The *Greek-based* culture of the Germans was lost, and in its place came an inferior *"nouvelle culture"* of the French Academy.

So what about their co-conspirators, the Dutch and English?

At the time, the Dutch were nouveau-rich traders of no great cultural development. At the time, the English were poor and isolated. Their great days were still ahead. By contrast, Paris was an intellectual and cultural powerhouse. And still is.

One could doubtless trace this still further back, to the failure of Charlemagne's sons to keep his kingdom united, to the conquest of Gaul — but not Germany — by Caesar, the Roman destruction of the preexisting Gallic culture, etc.

You want science? You want Germany. German engineering. German precision.

You want culture? You want diplomacy? You want the French. The opera. The ballet. The fashion.

We can have both, but more than that. *Every country is unique.* The Age of Enlightenment, the great Encyclopaedia, was a French thing. It was a mistake to impose that on the world. Science, as we know it today, is no more than whims that change from decade to decade. *You don't believe me?* Read old science texts. In 30 years, present-day texts will sound just as dated.

Not only must we restore astrology. We must also restore Aristotle. We must restore the very foundations.

Venus is retrograde in Gemini, can you tell? No sweet words, only sour ones. — *May 22, 2012*

🕐 **Dorothy** *and her magic broomstick*

It is not enough to revive or re-establish astrology. Astrology has been continually revived since Raphael (aka R.C. Smith) started the *Prophetic Messenger* in London in late 1826. Which already ignores Ebenezer Sibley, James Wilson, J.M. Ashmand, S.A. Mackay and many others who were active at that time.

Subsequently, Zadkiel (R.C. Morrison), Luke Broughton, William Chaney and others kept astrology going until Alan Leo could revive it yet again in the early 1890's. Who passed the baton to the young Vivian Robson and Charles Carter, who kept things going up to the mid 1960's. By which time Dane Rudhyar was firmly established, which led to Liz Greene, after which came the Hellenistic revival, which takes us to the present day and this newsletter. Astrology has been very nearly revived to death.

And with every new generation we hear the same story: That Astrology will finally be accepted, is about to be accepted, by The Establishment Itself. That it will regain its rightful place in the world, will be shouted from the rooftops, will be taught in academia.

Fat chance. Kepler College was briefly an officially accredited four year school, but in the ensuing protest lost it official status and is now another group of well-meaning teachers. If it survives it will take its place alongside a handful of other distinguished but unofficial schools, London's Faculty of Astrological Studies chief among them.

Science has long been identified, correctly in my view, as astrology's stumbling block. Which led to efforts to "prove" astrology by scientific means. Which ultimately produced three presentations. The first was by J.H. Nelson sometime in the 1950's, which, strictly speaking (radio wave propagation), was not astrological. The second, in 1964, was by George McCormack (astrometeorology). The third, in the late 1970's, was by Michel Gauquelin (Gauquelin sectors, ascendant/midheaven). All three efforts fell on very deaf ears. Gauquelin in particular became the target of intense scorn. Subsequent computer-driven efforts in the 1980's to "prove" astrology were fruitless.

So when I came up with my own theoretical construct a year ago I did not for an instant think it would get a hearing. As George W. Bush, a fearless leader, once said, *"Fool me once . . ."*

Which made me curious about this thing called "science." As an expression of an ultimate, and ideal, it was clearly beyond reproach, but as actually practiced, by mortal men with egos, it was as deeply flawed as any other religion. Which in many ways it resembles.

I investigated the origins of modern science and found, not to my surprise, that it was an urban, French-led movement. Which explained its easy condemnation of "rural superstition," but did not explain why astrology, to say nothing of Aristotle, got tarred and feathered as well.

For that I had to turn to the Renaissance, to discover the French had, then as well as now, a strong authoritarian bias coupled with a disdain of real learning. Astrologers have been, in effect, battling the goose-pate-eating fools of the French Academy.

And it's not that "science" does not like *astrology*. As the official science community is structured, it doesn't much like *science,* either. This is because authoritarian minds tend to get stuck in authoritarianism. Evolution, for example, got stuck in Darwin's slow and steady trap, from which it has long struggled to escape. The best evidence now suggests that severe cataclysmic events trigger not only mass extinctions, but also rapid shifts in animal and plant evolution, but this viewpoint cannot be heard because of all the "scientists" reflexively shouting it down.

Well, you think. That will end when the defenders are all dead and the next generation succeeds them. After all, science has always been faddish in this way, as anyone may read in old issues of scientific magazines. Well, maybe. Maybe not. It depends.

Science being, in reality, a blunt instrument, you may be surprised to learn that a lot of the most wonderful, the most subtle, the most innovative things simply got *ignored.* Wholesale. Since post Aristotle science is a consensus affair (prove it and we'll believe), then if the majority of those posing as scientists could not easily comprehend, then you and your work will be ignored and subsequently lost. This is, of course, the very antithesis of "science" and the very essence of a clique.

Of the fate of Nicola Tesla and his many wonderful inventions, the world now knows.

Tesla, a man with a soft, easy to chew ego, got bullied by the likes of Thomas Edison, George Westinghouse and others. The world really is a brutal place and if you do not force yourself upon it, you and your work will be lost, but I repeat myself. Science will not come to your rescue, for it is a clique that does not care. Which, regardless of its many faults, was never true of religion.

Religion differs from science in that it has an external authority over which it ultimately has no control. Anyone may directly perceive the trinity of God, for example. Which, literally, can be found in the Sun *(God)*, Moon *(Goddess)* and the relationship between them *(Holy Spirit)*. That close. That real. That tangible.

The Greeks gave science strong underpinnings, but the Enlightened French threw them away in favor of whatever whimsy sounded good. The result has been an utter mess. Many of you will disagree. Do we not have many wonderful things, is life not much better now than ever before?

Indeed it is, but *not* because of science. Life is better because of an *economic system* that harnesses *human work* to produce lots and lots of *money*. The difference between the Medieval period, and the modern world, is one of *sheer money*. Medieval Europe was *poor*. We are not.

Rich societies need banks and it is the nature of banks to so monopolize money as to impoverish the population. (Andrew Jackson fought the banks 180 years ago.) Just as science did not create the industrial revolution, science neither caused the Wall Street collapse that has now impoverished us, nor will ride to our rescue. Evolution? Who cares! Global warming? Not if there's no economy to heat things up. Self-important angels that dance on the head of a pin? Now you've got the idea!

In one respect, Tesla was lucky. He was lucky to be associated with people who are still household names, which meant that eventually his work would be revived. But there are many who were not so fortunate.

In reading Luke Broughton's book, *The Elements of Astrology*, which I have just reprinted, I read again of John Keely, the late 19th century inventor of the "Keely motor." I had only run across his name once before, in H.P. Blavatsky's *Isis Unveiled*, where marvelous things were said of him and his engine. That it was capable of sheer disintegration. Keely built over a hundred various "machines."

Some of Keely's earlier machines have come to light. As best anyone now knows, these early devices are elaborate, complex water hammers. A water hammer is what happens when a flow of water in a pipe is abruptly shut off. It "bangs" or hammers. Keely, an engineer, not a scientist, exploited that hammering in amazing ways, allegedly producing microwaves by means of water. What his machines can do no one now knows. No one even has a clue.

But science does not care that Keely and his work was lost, just as science does not care that George McCormack and his work was lost, or that astrology got short shrift and was nearly lost. Science only cares about winners. The powerful. Not losers. Not whiners.

The Dorothy Society

The aloof, arrogant elitism of science needs an analogy to be fully appreciated. It needs Dorothy.

Through no fault of her own, a tornado has marooned Dorothy, a native Kansan, in a strange land. Let's put her somewhere near Amarillo, Texas. To get home she is to follow the yellow path of knowledge to meet the Wizard. Who is the Chief of all Scientists.

When she arrives, he promptly brushes her off. *Little lost girls are not our responsibility*, he huffs. *If you want my help, do something worthy. Bring me the head of, I mean, bring me the broomstick of the Wicked Witch of the West.* Which sounds an awful lot like, We will believe astrology when you astrologers prove it. The Wizard does not expect Dorothy to return. Science does not think astrology even exists, so it can't possibly be proven, so they never have to make good on their end of their own bargain.

The Wizard, being a wizard, could easily dispose of witches if he wanted. Science could just as easily do what I have done, take the signs of the Zodiac out of the sky and put them in the Earth, which is essentially child's play. By comparison, Newton's laws, Kepler's theorems, Einstein's relativity, are truly awesome and complex. What I have done, grounding astrology and putting it in the earth, in hindsight, is a simple, even obvious thing.

In reality, the Wizard *needs* witches. He needs them as bogeys. Be good or the Wicked Witch will get you!

Just as science *needs* astrology: Learn your math and science! You don't want to grow up to be an *astrologer!* Horrors!

And so, like frightened children, we obey.

When Dorothy came back with the magic broom, when Gauquelin, McCormack and Nelson each presented their proofs, the Wizard's response was outright dismissal. Go away! We don't want you!

And that was that. Dorothy and her comrades were left spluttering. The Magnificent Oz had spoken.

Whereupon it was Toto, the dog, who drew back the curtain to reveal an old and tired curmudgeon in a tiny room, manipulating the theatre with his levers and switches. It turned out the Wizard had no magic. **Nor does science.**

What we need, therefore, is not another revival, nor a theory per se, nor even an expose of scientific flaws.

What we need is a society to codify and promote this multi-pronged approach:

Revive astrology.

Give it a solid theory.

188 The Triple Witching Hour

Revive Greek philosophy as the essential, overall scientific framework.

Make Astro-Medicine the spearhead.

We have the magic broomstick of astrology. We must now learn to use it.

A successful society will break the back of the present scientific Mafia. The science that survives will be good and true and just and will last. What does not was never science to begin with.

Having conceptualized the problem and having provided a framework for its solution, I am the logical person to found and direct such a society, but I am not a good choice. A full moon, from Aquarius to Leo, from 9 to 3, has made me a wary recluse.

At this point I proposed a society with me leading it, but the effort came to nothing. — *May 29, 2012*

☉ Clint Eastwood

Ken Bowser has a new book out, touting the virtues of the Sidereal Zodiac. Siderealists say that their Zodiac is the "true zodiac," based on actual star positions, as best those positions can be worked out, as the sky unfortunately lacks a distinctive marker, which in this case has been determined to be the star Spica. Vedic astrologers, who are also Siderealists, have a number of slightly different starting points.

By contrast, the Tropical Zodiac, commonly used in the west, has no fixed position in relation to the stars. The Tropical Zodiac, as I have mentioned before, is based on the solstices and equinoxes.

So which Zodiac is right? Which works the best, in practice? Can *both* work? Can we sort this out? As I have my own theory of astrology, a showdown between Ken's astrology and mine might be of interest.

I was going to use one of the charts in Ken's new book, *An Introduction to Western Sidereal Astrology*, but a few days ago I had a request for Clint Eastwood. I'm an old-timer, I remember Clint all the way back to his days in the TV show Rawhide. I've always liked the guy, I think everyone does. I checked with AstroDataBank and discovered his birth time was beyond question, which made him ideal.

Clint Eastwood was born May 31, 1930, at 5:45 pm PST, in San Francisco, CA. We will start with his chart in Sidereal.

Siderealists don't read their charts the way that I read mine. Ken says Siderealists don't use houses. He says, *"Western siderealists maintain that all the house systems are over-rated and that the houses are the least important and least consistent astrological index.... The sidereal position is to accord the houses secondary importance and to emphasize the very old doctrines of angularity and dispositors.... Medieval astrologers, in particular, consistently emphasized the angles and angular houses."* (pg. 31)

Which sounds fair enough. It would appear that Ken would agree that the ruler of the ascendant is a key player in the chart. While Ken says Siderealists do not care for one house system over another, I have found in the past they are partial to Campanus houses, as I have seen them used by numerous siderealists, Ken among them. Campanus houses date from the 13th century, they were the first house system developed in Europe and I believe the first developed with the new Arabic number system, which was vastly more powerful than the previous Roman number system. I am not certain to what extent medieval Europe used Campanus houses, as opposed to the earlier (and cruder) Porphyry, or Regiomontanus, which was developed a century later and which was in widespread use by the 17th century, by William Lilly among others.

Campanus was revived in the 20th century by Charles Carter who, I presume, influenced Cyril Fagan, the founder of modern Sidereal astrology. Since Ken does not actually use houses in his book and as I do, I am going to set Clint's sidereal chart, as well as his tropical chart, both in Placidus houses. With preliminaries out of the way, let's begin.

SIDEREAL

Clint Eastwood has Libra rising, which makes him handsome, if not actually pretty. As a Libra rising, he is, on the one hand vacillating, shifting from one position to another, as well as constantly pitting one faction against another. Libra is the sign of manipulation, conflict and warfare, which is often unappreciated. Eastwood has a keen appreciation for art and beauty, for sheer color, the many films he has made all have a splendor about them, a pure beauty, which is widely admired.

Chart ruler, Venus, is in Gemini, where it gives Eastwood a fluency, a beauty, with words and speech and writing in general. His famous spaghetti westerns of the late 1960's were memorable for the fluency with which Eastwood delivered his lines. Venus strung between Jupiter and Pluto, from one end of Gemini to the other, adds both volume (Jupiter) and intensity (Pluto) to his communication. When Eastwood speaks, every beautiful word is important, every word has intense meaning.

Venus in the 8th house is lucky with money. People—women especially—want to give him money. Note that Venus disposes the Sun, and Mercury, both in Taurus in the 7th house. Eastwood has profited financially from his numerous marriages and many girlfriends. Multiple marriages because with Aries on the seventh house, Eastwood is headstrong and tends to rush into marriages and partnerships on whim. As the ruler of the 7th, Mars, is in the 12th house of his partner, he has more than once been manipulated into marriage, which only comes out much later. Mercury, which is very near the cusp of the 7th, means his wives/girlfriends tend to be young, and fleet of foot. Eastwood no sooner starts a relation-

Tropical vs: Sidereal: Clint Eastwood 191

ship than someone new comes along to catch his eye.

Moon in Cancer in the 9th house, intercepted, Eastwood keeps his religious beliefs to himself. These beliefs are deeply felt (Moon/Cancer) and tend to be traditional. If his beliefs were challenged, he would likely be defensive, but as interceptions hide the planets in them, Eastwood's emotions are rarely seen.

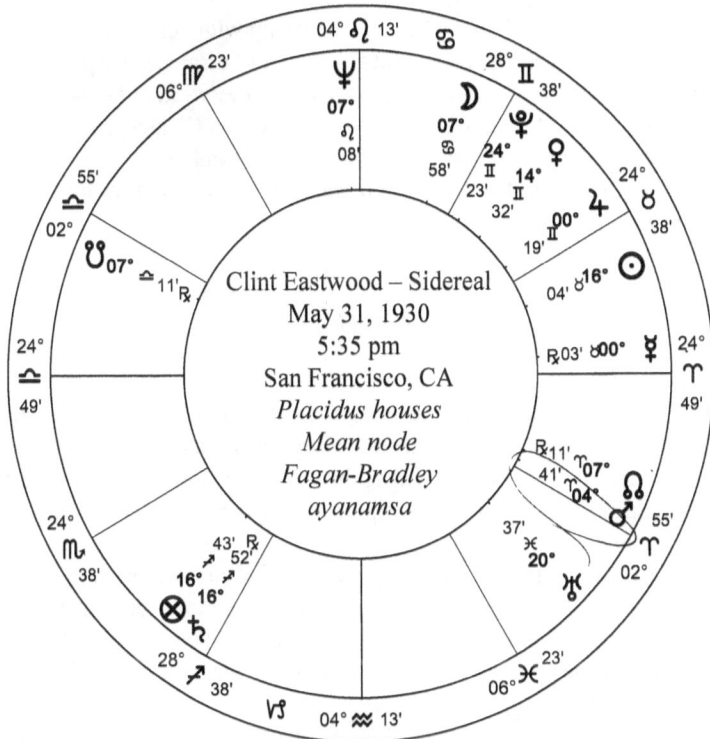

Clint Eastwood – Sidereal
May 31, 1930
5:35 pm
San Francisco, CA
*Placidus houses
Mean node
Fagan-Bradley
ayanamsa*

Saturn retrograde in Sagittarius very near the 3rd house cusp makes Clint travel whenever and wherever he can. Work this out with me: Third house is day-to-day life, Sagittarius wants to expand as much as it can, Saturn says no you don't, but as Saturn isn't actually in the 3rd house and as it's retrograde and doesn't want to go there, I "square the circle" by getting Clint on the road whenever he can. Note the fairy tight Saturn-Venus opposition. Venus, as chart ruler in Gemini makes Clint exceptionally well-spoken, but Saturn opposing means that what he says is frequently misunderstood, which gets him into trouble. Saturn ruling Capricorn, which is intercepted in the 3rd house, made his childhood long and difficult. Schooling was dull. The Moon in Cancer on the opposite side of the chart gave him no help. So far as the Moon is concerned, Clint would rather be daydreaming, imagining exotic homes (Cancer) far, far away. If the Moon was not intercepted, I would expect Clint to have found a home abroad.

Finally, the Midheaven. This is Leo, ruled by the Sun in Taurus, which is in the 7th house. Clint is known for being half of a celebrity couple. With Neptune there, the women he is seen with are always a surprise. We never know who he will turn up with next. Neptune in the midheaven, Sun as the ruler, Clint is a master of disguise. He prides himself (Leo) on slipping in and out of public view, without being noticed. This is also a feature of his films, where, like Lon Chaney, he frequently dons character makeup. A new "man of a thousand faces."

Such is Clint Eastwood, **Sidereal**. In your opinion, is this the true man, YES or NO?

TROPICAL
A life in pictures

With Clint Eastwood's chart in the Tropical zodiac, we start again at the ascendant.

In his Tropical chart, Scorpio rises. Scorpio is intense and secretive. It believes in SILENCE. It is a sign of life and death, of absolute power. Before I go further I will again quote Ken Bowser, the Siderealist: *The sidereal position is . . . to emphasize the very old doctrines of angularity and dispositors.*

Scorpio is ruled by Mars, which we find in Aries in the 6th. The ruling planet in its own sign reinforces the strength of the ascendant as well as adding the ruling planet's other sign to it. This is a nativity which is sneaky, which will use force in subtle and masterful ways, and which is not afraid of sheer hot-headed rage.

Mars in the 6th is a sign of tireless work, in Aries of eagerly taking on projects, which overwhelm his co-workers. With its rulership of the first, we may expect Eastwood to be in superb physical shape. Which I hear he actually is.

In mundane charts, which are the charts of nations, or the various ingress charts, the 6th house is the house of the army and police. In a nation, the army and police have exclusive use of coercive force, in other words, the use of firearms. Clint Eastwood's movie career spans more than 50 years. Presuming he has tended to play himself, Mars ruling his Scorpio ascendant, from Aries in the 6th house, aptly describes a *silent loner* (Scorpio) who is believed to have *immense power* (Scorpio), who uses it in *sudden* and *abrupt* ways (Mars in Aries) by means of firearms (6th). The list of Eastwood's movies where he plays the lead with these

Tropical vs: Sidereal: Clint Eastwood

precise characteristics is quite long, as we all know. What little dialogue Eastwood has in his movies is terse and abrupt. *"You feel lucky, punk? Well do ya?"*

It is also a characteristic of Mars in the 6th, if it is in any way comfortable there, that the native will never retire, that he will work until he drops dead. Mars in the 6th expresses itself through work. Eastwood has just turned 82 and shows no sign of stopping.

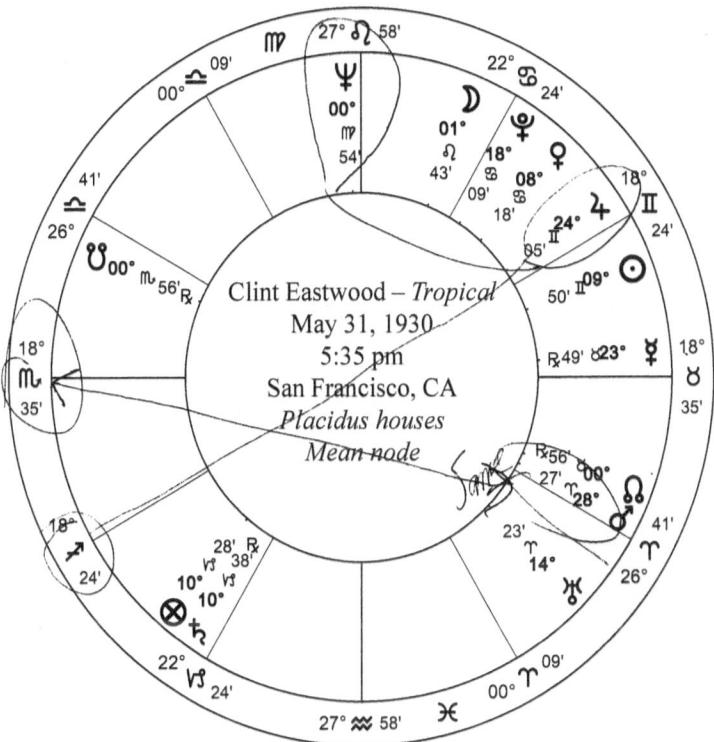

Romantically, Eastwood likes them young and pretty. Pretty, because Taurus is on the cusp of the 7th. Young, because Mercury is there. As with the Sidereal analysis, Eastwood's women are fleet of foot. The list of women he has been associated with is quite long. He has fathered seven children by five women.

At the stated time of 5:35 pm, Eastwood does not have an intensely fertile chart. Fertility is shown, first by the sign on the cusp of the 5th. Fertile signs, such as Cancer, Scorpio and Pisces, all but guarantee children. Fire signs, such as Aries, which we find on the 5th cusp in Clint's chart, all but exclude them.

Next we look at the ruler of the 5th, which we find to be Mars in Aries. Which is not good.

Followed by the Moon, which we find in Leo. Which is not fertile.

Then we look at the ascendant. Which, in Scorpio, is as wet as it gets. We look at Eastwood, we think, this is one sexy, virile man. Sexy he is. Virile he is not.

Then we flip the chart and look at it from the point of view of his females. Eastwood's 11th house is his girlfriends' 5th house. It's the house of the girlfriend's children.

The 11th house has Libra on the cusp. Libra is moderately fertile. It is ruled by Venus in Cancer, which is hugely, desperately, immediately fertile. Note also that Venus is placed in the second house of the partner. So far as Eastwood is concerned, his seven children are not his, were never his. He in fact has repeatedly denied paternity. These are not his children, but the children of his girlfriends. In Eastwood's opinion, his empty 5th house means he has no children. His empty 11th house means, so far as he's concerned, that his girlfriends have none, either. Astrology is just that explicit.

What about Uranus in the 5th? Uranus means the children were surprises. Uranus did not make them unwanted and Uranus did not make Eastwood deny them. Uranus in this case is a trivial detail.

In Eastwood's chart, Mercury and the 7th cusp are ruled by Venus in Cancer in the 8th. Cancer in the 8th is a very personal placement. One *possesses* money. Because it's Eastwood's 8th house, he is of the opinion that it's *his money*. Because the 8th house is his partner's house of money, and because Venus rules the 7th from the 8th, the partner will think the money is *hers*. As the rulerships give his girlfriends control over the money, the result, in fact, has been a messy divorce and at least one out of court "palimony" settlement. Eastwood paid $25 million for a divorce in 1984, along with an undisclosed sum to a former girlfriend in 1999.

Note that Eastwood's 8th house money problems are compounded by his second house, which has Sagittarius on the cusp. The house with Sagittarian on the cusp is the house of *hopes* and *desires*. In this case, Eastwood hopes and desires to have lots and lots of money, but lacking any planet in the second this is mostly a dream. Sagittarius says Clint might earn his money by some means of communication that is broadly based, that is not aimed at any single individual, individuals being Gemini.

And we look across from the 2nd to the 8th and find Jupiter, debilitated in Gemini. Which, firstly, confirms our hunch that Eastwood would use some means of communication to make lots and lots (Jupiter) of money, which, as debilitated planets want to be in the house opposite, Jupiter would consider to be its own. Jupiter debilitated, the actual vocal communication in those movies would be sharply limited, as Jupiter does not know how to make small talk. (That's Mercury's job.) When we look at Eastwood's most notable characters — Dirty Harry, the Man with No Name, etc., we are torn between thinking he does not *want* to speak, or perhaps does not quite *know how.*

The result is that Eastwood has only a tenuous hold on his money, and has been forced to give large amounts of it to the women he has known.

Which amounts to a forecast. You see a chart like Eastwood's, you say this man will fight over money and end up paying a lot of it to wives and girlfriends. Forecasters will look at a prediction like this, take it out of context, and wonder how transiting Saturn on Eastwood's ascendant, along with Uranus making a station in the vicinity of Eastwood's 2nd house cusp (1984), or Saturn on Eastwood's descendant (1999) would result in such drastic financial outcomes. It is the natal chart itself that must guide you. Every chart has trap doors in it. Severe transits will set them off.

There was a famous forecast, by Gaurico, that King Henry II, of France, would die by being gored through his right eye in a tournament. Which, amazingly enough, came to pass in 1559. Modern astrologers have long been puzzled how such a specific forecast was made. It was simple. Henry's exposed Sun was hit by (I presume) an eclipse in such a way as to focus attention on the king's right eye, which the Sun rules. The eclipse came to pass and thereafter one simply awaited the trigger that would bring the horrific event to fruition. (This is the simplest way in which the prediction could have been made.) It is the same with Eastwood and his money and his women. Should Clint Eastwood still be with us in 2014, he might suffer the same economic loss for the third time.

Note also that Saturn and Venus are opposed, Capricorn to Cancer. As Saturn rules Capricorn, it is inherently stronger than Venus in Cancer, but as Saturn is retrograde, Eastwood does not quite know how to handle Saturn's insecurities. With Saturn ruling Capricorn and Mars as chart ruler and in Aries, this is a powerfully Mars-Saturn man, and, with Venus on the other end, we have a classic cardinal-sign T-square. T-squares in cardinals are endlessly dynamic, never more so when the planets that compose it inherently hate each other, as Saturn, Mars and Venus do. We note that while Mars, well-placed in Aries, likes Capricorn, both Saturn and Venus positively hate Aries. And that while Mars might like Venus, he hates Cancer, the sign she is in.

So not only are the planets in Eastwood's T-square antagonistic, the planets make the signs themselves antagonistic, and the aspects — two squares and an opposition — are also antagonistic. You will say that Mars is not really in orb, but this is a judgement call. Your second cousin three times removed is not a close relative, but if you married her and it didn't work out, the blood tie will be an aggravating factor.

Saturn opposite the ruler of his 7th house makes Clint naturally suspicious of the women he is with. All the more so as time passes, as Saturn himself is the planet of age. Remember that old men with young women become increasingly suspicious with every passing year, and, I fear, with reason. When the opposition is stressed, it is Mars, in square to both, and as chart ruler, who will lash out to resolve matters. In this matter, note the autobiography by Sondra Locke, *The Good, the Bad and the Very Ugly*, a memoir of her time with Mr. Eastwood.

In Eastwood's Tropical chart, Saturn is firmly in the 3rd house, in large part because in the *Tropical* zodiac it is in the same sign as the cusp itself. Here, it shows a long, hard onerous childhood, which in fact was spent in the depths of the Depression. Retrograde, Saturn, aka Eastwood, never wants to go back there. With his Part of Fortune exactly there, he regards his childhood as distinctly unlucky, Saturn's retrograde condition spoiling it. On the other hand, Saturn keeps Eastwood from ever quite breaking free of his childhood. He grew up in San Francisco. For many years he has lived a short way down the road, in Carmel.

Note carefully what I've done. I've put Eastwood's Saturn in his 3rd house because of simple sign affinity. The orb, from Saturn to the 3rd cusp, is 12 degrees.

I've put his Venus in his 8th, even though its sign, Cancer, does not match the sign on the 8th itself, which is Gemini, and that it does match the sign on the 9th, the orb being 14 degrees. Why am I certain of Venus being in the 8th?

I put Venus in the 8th firstly because of the position of Pluto. Pluto at 18 Cancer effectively "cuts off" Venus's attempts to associate with the 9th house cusp. When a planet has to "crawl over" some other planet to "get into" the house it is otherwise associated with, it just won't. It makes no difference that Pluto is a "nasty" planet. Mercury, even direct, is just as much a barrier.

And I consider Venus to be in Eastwood's 8th house because it makes for a better story.

Women fighting Eastwood for his money, and he being forced to part with a great deal of it, makes sense.

So what to do with Pluto? Essentially, I've ignored it. Eastwood's Pluto is in the 9th, yes, I'm okay with that (his Iwo Jima movie has Pluto in the 9th all over it), but far more powerfully, Eastwood has his Moon in Leo in the 9th. Moon in Leo in the 9th are powerful, firmly held beliefs. Beliefs with which he identifies utterly. Beliefs which, as Leo is a fixed sign, never change. When the Moon is in Leo the native has a sense of himself as unique and singular, but as the Moon, unlike the Sun, is vari-

Tropical vs: Sidereal: Clint Eastwood   197

able, it only sometimes wants the spotlight, in fact it can only stand the spotlight if taken in measured doses.

Finally, note the position of Neptune, in the 10th. Note that in the Tropical chart, Neptune is not in the same sign as the MC itself. In Eastwood we see a man, Leo on the MC, who is unique and an individual, one who is proud (Leo) of himself, but, Neptune nearby in Virgo, our actual image of him is clouded, the details are not clear. We want him to be larger than life, but when we study the man (Virgo, the sign of study), we are taken immediately back to his wives and girlfriends, as the ruler of Virgo is Mercury, which is in the 7th, and an entirely new story opens up in front of us.

One that, you will note, Eastwood has never explored, not in any of his many movies. Given the innate power of Mars as chart ruler, and the Saturn-Mars-Venus T-square, the story of Eastwood's life is one that few will risk telling, at least as long as Eastwood is alive. Sondra Locke's memoir would make an excellent movie but you will note that Hollywood has not touched it, despite its obvious blockbuster status. People fear this man.

So how do Tropical and Sidereal stack up?

Presuming the basis of our analysis is correct, which is that **when the sign on the ascendant changes, the chart ruler changes,** and that when the chart ruler changes, there are key house and sign changes as well, *presuming this is true and correct,* the Sidereal Zodiac, *as a means of reading a natal chart,* appears to be of little use.

But I am still trying to save Ken Bowser's premise, that angles and their rulers are to be our primary guide to chart interpretation. I am reading his delineations and noting that he does not actually use dispositors. Presumably he does not know how. For the most part, Ken uses aspects. Aspects by themselves have no context — *none whatever.* Aspects merely show that one planet will relate to another. Like one of your in-laws will relate to another of your in-laws.

You do not know what Mars will do with Saturn until you put them in a context. No, they don't much like each other, but exactly what they will do has many variables. It is a failing of modern astrologers to project what they like upon aspects, to make Saturn to be daddy, or maybe that's the Sun who is daddy, and the Moon to be mommy, etc. Incessant repetition of these mantras, some book-length, without regard for signs or houses, is not only meaningless, but when aspects are over-delineated in this way, *those delineations stop us from getting further into the chart itself.* When Saturn is always daddy, then daddy is a stick-figure, a one-dimensional tyrant, the same daddy for you, the same daddy for me, an imaginary daddy that bears only a casual relation to the real man we grew up with.

It's when we make daddy a *house* and then seek out the *planet* that *rules* that *house,* that we *understand daddy.* Eastwood's daddy, for example, was a quirky, hard-working no-show. How do I know that? Simple. Daddy is the 4th house. The fourth house in Eastwood's chart is *Aquarius* (quirky). Saturn is in *Capricorn* (hard working), and in the *12th house from itself* (the third house is the 12th house from the 4th), so Clint's daddy was never around. That Daddy Eastwood tried to be a father is shown by Saturn's retrograde condition. He did not want to be hidden in the 12th, he wanted to move to the 11th and be Clint's friend, but circumstance (the struggle for work: *Capricorn)* would not let him. You can find endless such details in a chart.

Eastwood's mother is shown by the 10th house, where we find Leo. Leo means she wore the pants in the family. She wanted to be a genuine friend to Clint, to talk to him, as her ruler, the Sun, is in her 11th house and in Gemini. Did she succeed? Probably not. Eastwood's Sun is largely stranded, and in Gemini she told him stories. Which her Scorpion son saw through in an instant. Why did I not put these notes under the Tropical section? Because Eastwood's parents don't seem to be a big deal in his life.

In Eastwood's case, the difference between Tropical and Sidereal is the difference between a *sweet-talking dandy who lives off women* (Sidereal), and a *mean, nasty s.o.b. with powerful emotions* (Tropical). We are forced to this difference simply from considering the different rising signs and their ruling planets. Eastwood is either LIBRA/VENUS or SCORPIO/MARS. We are confronted with this choice at the very beginning of our analysis. He *cannot* be both Venus *and* Mars. He *must* be *one* or the *other.* We must decide, and our decision will determine our zodiac.

When we look further, when we look at the rulers of the angles, when we look at the Sun, its Sidereal and Tropical signs and house, the Moon and its signs and house, when we consider the two sets of houses, when we note the intercepted signs and how they shift from Tropical (4th-10th) to Sidereal (3rd-9th), when we consider all these factors, it's frankly not even close. Tropical wins, hands down.

So *(sooooooooo),* why is there a Sidereal Zodiac? At all?

The long-standing belief has been that the signs of the Zodiac, both Tropical and Sidereal, are external to the Earth, that they fall to the ground as energies from sources in the very stars themselves.

There is no evidence, *no evidence of any kind,* to support the belief that the zodiac falls from the sky. I confess it took me years of hammering to stumble upon the simple fact that *the zodiac is of the Earth,* but once I realized the truth, a thousand of years of illusion disappeared in an instant.

The Zodiac comes from the stars, yes? Then why are all the signs of equal strength? When, in fact, some signs have lots of big powerful *first magnitude stars,* while others have *no powerful stars at all?*

The Zodiac comes from arrangements of stars known as constellations, yes? Then why, despite the fact that each constellation varies in size, both in longitude as well as latitude, do we commonly chop them into precise 30° segments that, in width, all exactly span the tropics?

Why, if the Zodiac is external to the Earth itself, a Zodiac which comes from a source (or sources) far, far, far away, do we impose so many local conditions on it? How can we impose limitations on what is external to us?

How can it be that after so many thousands of years that fakes and frauds and pretenders can even dare to claim there are more or less than a dozen signs which compose it?

All of Astrology is explained when we put the signs of the zodiac in the Earth itself. If there is in fact a six-sided crystal at the center of the Earth, then its properties will describe astrology precisely. Presuming all the other planets are also enrobed crystals, **astrology is nothing more than the study of planetary crystals.**

Sidereal Astrology, one of a number of astrological schools, works, or seems to work, as long as we use a crippled aspect-based system. So long as we don't look at the ruler of the ascendant, so long as we abandon houses and ignore dispositors, so long as we fail to grasp the essential sign-house-planet trilogy, so long as we are satisfied with vague generalities. Judging by the enthusiastic response I get to these newsletters, many of you find my house-based system to be a revelation.

In its favor, Sidereal astrology, which *is not based on stars at all,* but on the **Earth's wobble,** is **excellent for predictive work.** This is

because the period of the Earth's wobble, the time it takes for one full cycle, around 25,800 years, is the ultimate chronocrator, the ultimate time marker.

My conclusion: **Use the Tropical Zodiac for natal delineation. Use the Sidereal Zodiac for forecasting.** *Use what works.* Use both. Why not? Why limit yourself?

As for myself, remember that I analyze. I do not forecast, as I do not understand it. So I will naturally gravitate to the Tropical. Everyone is limited in this way, everyone is limited in one way or another. — *June 5, 2012*

☉ Fix Your Heart With Astrology and Herbs

This is from Blagrave's *Astrological Practice of Physick*, from the book I restored, reset and published two years ago. I knew how it worked at the time but until this past month, I had never tried it:—

You take three herbs that are good for the heart, harvest them during a planetary hour ruled by the Sun (this is *required)*, wrap them in a paper towel, tie it up with string, wear it around your neck, and it makes your heart stronger. I was too lazy to actually grow the herbs.

But this past spring my heart was so very weak that I got scared. So I went to Blagrave and read his list of herbs for the heart:

Angelica, marigolds, borage, balm, rosemary, bayberries, costmary burnet, cinnamon, cloves, endive, sage, saffron, nutmeg, strawberries, damask roses, spikenard, galingale, hart's tongue, lavender, sandalwood, vipers grass.

Of these, marigolds, lavender, sage and rosemary can be had, as living plants, at any nursery or home improvement store. I got some, I waited for a good moment to plant them and then I waited, nervously, until they were strong enough for clippings.

And then on a Sunday morning at sunrise two weeks ago (hour of the Sun), I did my first harvest: *Marigold, lavender, sage.* Wrapped them in a paper towel and put them around my neck and kept them there, 24/7.

Didn't feel any different. The next Thursday I had to mow the lawn, which is a riding lawn mower, followed by an area where I use a conventional push mower. Which for the past year had left me sitting flat on the ground, gasping for breath, feeling as if I was going to die. But I got all the way through it and was hardly winded. I was amazed.

The next day, Friday, was my weekly acupuncture session. The doctor commented how good I looked. I was surprised again. On Sunday at sunrise I again got up, went down to the herbs, and replaced the herbs with new ones: Rosemary, lavender, sage. Yesterday I replaced them again: Marigold, lavender, sage. At the start of the third week, I feel

Fix Your Heart!

much stronger.

Blagrave says to do this, not just for weak hearts, but for all ailments, because a strong heart makes everything better. A strong heart is the definition of good health. The formula is simple:

First, grow herbs that are good for the heart. Pick from the list. Yes. The little potted plants at Home Depot, WalMart or Lowes will do just fine. Transfer them into a window box when the moon is waxing and in a water sign, which will be June 29-30. If you can't wait that long, buy extra plants and take clippings directly from them.

Here's the important part: **You must harvest during the planetary hour of the Sun.** The easy way to do this is at **Sunday** at **sunrise**, since Sunday is the day of the Sun and the first hour after sunrise is the hour of the Sun. You can get the time of sunrise from your local paper, or from the Weather Channel's *Local on the Eights* (most of them), or by setting a chart for today with the Sun exactly on the ascendant, or you can Google *Time of Sunrise*.

Sunrise is a local event. You must get sunrise for the town you live in. Since sunrise changes according to the day of the year, and the length of the planetary hours change according to date and your longitude north or south of the Equator, as a rule of thumb **you are safe to harvest up to half an hour after sunrise**, winter, summer, spring or fall, anywhere from 60° north to 60° south. Snipping plants takes only seconds.

For other Sun hours on other days you will need to calculate: Take the times of sunrise and sunset, find the number of minutes from rise to set, divide by 12 and you have the length of one planetary hour, etc. Or download **ChronosXP** and it will keep track of it for you. (The download is sticky, I had to try three times.) When my plants get stronger I want to renew the herbs twice a week, on Wednesday or Thursday.

With careful selection of heart-herbs, this is foolproof. (DO NOT put other plants or herbs, including those ruled by the other planets, anywhere near your heart!) As it is non-invasive, three heart-herbs in a bag will work for all people, and for those under a doctor's care, it will work with all methods of treatment. It is easy and cheap and the results are wonderful. I am surprised and impressed. I wish I had done this years ago. — *June 5, 2012.*

Update, July 26, 2012: I've come to have a preference for marigold, sage and lavender. It's been nearly two months, I would not be without my satchel of herbs.

Bibliography

Books mentioned in this book. Some mentions are oblique:—
Abu 'Ali Al-Khayyat (*trans:* James Holden) *The Judgments of Nativities*, AFA, 2009
Al Biruni (trans: R. Ramsay Wright) *The Book of Instruction*, Astrology Classics, 2006
Anrias, David, *Man and the Zodiac*, Astrology Classics, 2010
Appleby, Derek, *Horary Astrology*, Astrology Classics, 2005
Bergeron, Tom, *I'm Hosting As Fast As I Can!*, Harper 1, 2009
Blagrave, Joseph, *Astrological Practice of Physick*, Astrology Classics, 2010
Blavatsky, Helena, *Isis Unveiled*, TPH, 1877
Bowser, Kenneth, *An Introduction to Western Sidereal Astrology*, AFA, 2012
Broughton, Luke, *The Elements of Astrology*, Astrology Classics, 2012
Patty Tobin Brittain, *Planetary Powers: The Morin Method*, AFA, 2011
Campion, Nicholas, *The Book of World Horoscopes*, Cinnabar Books, 1999
Carter, Charles, *An Encyclopædia of Psychological Astrology*, Astrology Classics, 2003
Cornell, H.L., *The Encyclopædia of Medical Astrology*, Astrology Classics, 2003
Culpeper, Nicholas, *Astrological Judgement of Diseases from the Decumbiture of the Sick,* and, *Urinalia*, Astrology Classics, 2003
Davison, Ronald, *Synastry*, Aurora Press, 1983
deVore, Nicholas, *Encyclopedia of Astrology*, Astrology Classics, 2005
Dorotheus of Sidon (trans: David Pingree), *Carmen Astrologicum*, Astrology Classics, 2005
Dykes, Ben (translator), *The Book of the Nine Judges*, Cazimi Press, 2011
Frawley, John, *The Horary Textbook*, Apprentice Books, 2005
Frawley, John, *Sports Astrology*, Apprentice Books, 2007
Goldstein-Jacobson, Ivy, *Simplified Horary Astrology*, 1960
Grove, George, *Beethoven and His Nine Symphonies*, Dover, 1962
Hamaker-Zondag, Karen, *The House Connection*, Samuel Weiser, 1994
Holden, James H. and Hughes, Robert A., *Astrological Pioneers of America*, AFA, 1988
Lilly, William, *Christian Astrology, Books 1 and 2*, Astrology Classics, 2004
Lilly, William, *Christian Astrology, Book 3,* Astrology Classics, 2005
Locke, Sondra, *The Good, the Bad and the Very Ugly,* William Morrow, 1997

Bibliography 203

Louis, Anthony, *Horary Astrology Plain and Simple*, Llewellyn, 1998
Maison Rosicrucienne, editeur, *The Rosicrucian Ephemerides, 1900-2000*, Rosicrucian Fellowship, 1983
McCormack, George J., *A Text-Book of Long Range Weather Forecasting*, Astrology Classics, 2012
Michelsen, Neil F., *Tables of Planetary Phenomena, 3rd edition*, Starcrafts Publishing, 2007
Morin, Jean-Baptiste (*trans:* Richard Baldwin), *Astrologia Gallica Book 21*, AFA, 1974
Morin, Jean-Baptiste (*trans:* J.H. Holden), *Astrologia Gallica Book 26*, AFA, 2010
Mull, Carol, *750 Over the Counter Stocks*, AFA, 1986
Oken, Alan, *Alan Oken's Complete Astrology*, Bantam, 1980
Robson, Vivian, *Electional Astrology*, Astrology Classics, 2005
Robson, Vivian, *The Fixed Stars and Constellations in Astrology*, Astrology Classics, 2005
Robson, Vivian, *A Student's Text-Book of Astrology*, Astrology Classics, 2010
Rodden, Lois M., *Mercury Method of Chart Comparison*, AFA, 1993
Roell, David, *Duels At Dawn*, Astrology Classics, 2012
Roell, David, *Skeet Shooting for Astrologers*, Astrology Classics, 2011
Rosicrucian Fellowship, *Tables of Houses*, Oceanside CA, 1995
Sakoian and Acker, *The Astrologer's Handbook*, Quill, 2001
Sargent, Lois Haines, *How to Handle Your Human Relations*, AFA, 2006
Saunders, Richard, *The Astrological Judgement and Practice of Physick*, Astrology Classics, 2003
Schmidt-Gorg, Joseph, and Schmidt, Hans, *Ludwig van Beethoven Bicentennial Edition 1770-1970*, Beethoven-Archiv Bonn and Deutsche Grammophone Gesellschaft MBH-Hamburg, 1970
Selby, Agnes, *Constanze, Mozart's Beloved*, Turton & Armstrong Pty. Ltd, 1999
Smith, William Henry, *Smith's Family Physician*, Hunter, Rose, 1873
TenDam, Hans, *Exploring Reincarnation*, Penguin Arkana, 1990
Tyl, Noel, *Astrology of the Famed*, Llewellyn, 1996
Vettius Valens (trans: Andrea Gerhz), *Anthology, Book 1*, Moira Press, 2011
Vettius Valens (trans: Mark Riley), *Anthologies*, Astrology Classics, 2013(?)
Wegler, Franz, and Ries, Ferdinand (trans. Frederick Noonan), *Beethoven Remembered*, Great Ocean Publishers, 1987
Wheelan, Joseph, *Jefferson's War, America's First War on Terror, 1801-1805*, Public Affairs, 2004

About the Author

David Roell was born in Kansas, of Steele County, Minnesota migrant parents. He was raised in Wichita, Hutchinson, Wellington, Winfield, Douglas, El Dorado, Meade, Iola, Greensburg, Salina and other communities. He has brothers and sisters. In Meade and Iola they lived in houses. The family vacationed in Dodge City, St. John, Owatonna and Austin. Roell attended community college, he graduated from University. Twice. He once spoke French.

For several years he lived in Europe: London, Paris, Montpellier, Strasbourg, with holidays in Calais and Barcelona. In the U.S. he has lived in New York, Los Angeles, New York, Los Angeles, New York, Ventura, Santa Fe, New York and Maryland, with vacations in Cozumel, San Francisco and Florida. Vacations are important.

He began his study of astrology at Broadway and 8th Street in New York, on October 24, 1983 at approximately 4:00 pm. He has studied intensely since the late 1990's.

From 1986 to 1990, he worked at the New York Astrology Center.

In August 1993 he founded The Astrology Center of America. AstroAmerica.com dates from January, 1996. Astrology Classics started in 2002. The newsletter is from July, 2007, his first book of essays, *Skeet Shooting for Astrologers*, was published in February, 2011.

Roell is married and has a daughter.

Better books make better astrologers.
Here are some of our other titles:

AstroAmerica's Daily Ephemeris, 2010-2020
AstroAmerica's Daily Ephemeris, 2000-2020
- both for Midnight. Compiled and formatted by David R. Roell

Al Biruni: **The Book of Instructions in the Elements of the Art of Astrology**, 1029 AD, translated by R. Ramsay Wright

David Anrias: **Man and the Zodiac**

Derek Appleby: **Horary Astrology: The Art of Astrological Divination**

E.H. Bailey: **The Prenatal Epoch**

Joseph Blagrave: **Astrological Practice of Physick**

Luke Broughton: **The Elements of Astrology, 1898**

C.E.O. Carter:
The Astrology of Accidents
An Encyclopaedia of Psychological Astrology
Essays on the Foundations of Astrology
The Principles of Astrology, Intermediate no. 1
Some Principles of Horoscopic Delineation, Intermediate no. 2
Symbolic Directions in Modern Astrology
The Zodiac and the Soul

Charubel and Sepharial: **Degrees of the Zodiac Symbolized**, 1898

H.L. Cornell: **Encyclopaedia of Medical Astrology**

Nicholas Culpeper: **Astrological Judgement of Diseases from the Decumbiture of the Sick**, 1655, *and,* **Urinalia**, 1658

Dorotheus of Sidon: **Carmen Astrologicum**, c. 50 AD, translated by David Pingree

Nicholas deVore: **Encyclopedia of Astrology**

Firmicus Maternus: **Ancient Astrology Theory and Practice: Matheseos Libri VIII**, c. 350 AD, translated by Jean Rhys Bram

Margaret Hone: **The Modern Text-Book of Astrology**

Alan Leo:
The Progressed Horoscope, 1905
The Key to Your Own Nativity, 1910
Dictionary of Astrology, edited by Vivian Robson, 1929

William Lilly
Christian Astrology, books 1 and 2, 1647
The Introduction to Astrology, Resolution of all manner of questions.
Christian Astrology, book 3, 1647
Easie and plaine method teaching how to judge upon nativities.

George J. McCormack: **A Text-Book of Long Range Weather Forecasting**
With Foreword by David R. Roell, Astrology At Our Feet

Jean-Baptiste Morin: **The Cabal of the Twelve Houses Astrological**
translated by George Wharton, edited by D.R. Roell

Claudius Ptolemy: **Tetrabiblos**, c. 140 AD, translated by J.M. Ashmand

Vivian Robson:
Astrology and Sex
Electional Astrology
Fixed Stars and Constellations in Astrology
A Beginner's Guide to Practical Astrology
A Student's Text-Book of Astrology, Vivian Robson Memorial Edition

Diana Roche: **The Sabian Symbols, A Screen of Prophecy**

David Roell:
Skeet Shooting for Astrologers
Duels At Dawn, the second book of essays

Richard Saunders: **The Astrological Judgement and Practice of Physick**, 1677

Sepharial:
The Manual of Astrology, the Standard Work
Primary Directions, a definitive study
Sepharial On Money. In one volume, complete texts:
- **Law of Values**
- **Silver Key**
- **Arcana, or Stock and Share Key** — first time in print!

Zane Stein: **Essence and Application, A View from Chiron**

James Wilson, Esq.: **Dictionary of Astrology**

H.S. Green, Raphael and C.E.O. Carter
Mundane Astrology: 3 Books, complete in one volume.

If not available from your local bookseller, order directly from:
The Astrology Center of America
207 Victory Lane
Bel Air, MD 21014

on the web at:
http://www.astroamerica.com

www.ingramcontent.com/pod-product-compliance
Lightning Source LLC
Chambersburg PA
CBHW030139170426
43199CB00008B/126